Adaptive Search and the Management of Logistics Systems

OPERATIONS RESEARCH/COMPUTER SCIENCE INTERFACES SERIES

Series Editors

Professor Ramesh Sharda
Oklahoma State University

Prof. Dr. Stefan Voß
Technische Universität Braunschweig

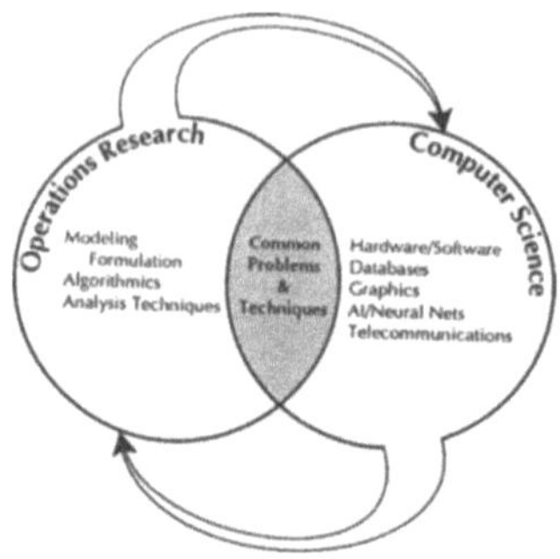

Other published titles in the series:

Brown, Donald/Scherer, William T.
Intelligent Scheduling Systems

Nash, Stephen G./Sofer, Ariela
The Impact of Emerging Technologies on Computer Science and Operations Research

Barth, Peter
Logic-Based 0-1 Constraint Programming

Jones, Christopher V.
Visualization and Optimization

Barr, Richard S./ Helgason, Richard V./ Kennington, Jeffery L.
Interfaces in Computer Science and Operations Research: Advances in Metaheuristics, Optimization, and Stochastic Modeling Technologies

Ellacott, Stephen W./ Mason, John C./ Anderson, Iain J.
Mathematics of Neural Networks: Models, Algorithms & Applications

Woodruff, David L.
Advances in Computational and Stochastic Optimization, Logic Programming, and Heuristic Search

Klein, Robert
Scheduling of Resource-Constrained Projects

ADAPTIVE SEARCH AND THE MANAGEMENT OF LOGISTICS SYSTEMS
Base Models for Learning Agents

CHRISTIAN BIERWIRTH
University of Bremen, Germany

Springer Science+Business Media, LLC

Library of Congress Cataloging-in-Publication

Bierwirth, Christian.
Adaptive search and the management of logistics systems : base models for learning agents / Christian Bierwirth.
p. cm. -- (Operations research/computer science interfaces series / ORCS 11)
Includes bibliographical references and index.
ISBN 978-1-4613-4679-1 ISBN 978-1-4419-8742-6 (eBook)
DOI 10.1007/978-1-4419-8742-6
1. Production management--Mathematical models. 2. Genetic algorithms. 3. Business logistics. I. Title. II. Series.

TS155.B495 1999
658.5'1--dc21

99-048109

Originally published by Kluwer Academic Publishers **in** 2000
Softcover reprint of the hardcover 1st edition 2000

Printed on acid-free paper.

Contents

Preface

Global competition and growing costumer expectations force industrial enterprises to reorganize their business processes and to support cost-effective customer services. Realizing the potential savings to be gained by exacting customer-delivery processes, logistics is currently subject to incisive changes. This upheaval aims at making competitive advantage from logistic services instead of viewing them simply as business necessity. With respect to this focus logistics management comprises

> the process of planning, implementing, and controlling the efficient, effective flow and storage of goods and services, and related information from point of origin to point of consumption for the purpose of conforming customer requirements[1].

This definition implies a holistic view on the logistic network, where the actors are suppliers, manufacturers, stockkeepers, shipping agents, distributors, retailers and finally consumers. The flow of goods along the supply chain considers raw-materials, work-in-process parts, intermediate and finished products, and possibly waste. The prevailing management of logistics operation is driven by aggregated forecasting of these material flows.

Modern logistics management propagates a disaggregated view of the material flow in order to meet the precise requirements at the interface between actors in the supply chain. Replacing aggregated information by detailed values establishes the prerequisites for an integrated process planning which goes for the shift from *anticipatory* towards *response-based logistics*[2]. Smaller units of goods are considered at shorter periods for planning, implementing and controlling the material flow. From

[1] cf. the Council of Logistics Management (1995).
[2] cf. Bowersox (1999).

the detached view of an actor this demands a decentralized process re-planning in order to adjust its daily logistics operation. From an overall view, bounded anticipation and reactive behavior of actors are locally combined which is assumed to improve the overall operation in the entire network.

Due to great advances in communication and transportation technologies, detailed information of material flows can often be exchanged immediately between the actors in a supply chain. This is undoubtedly the most important source contributing to the design of modern logistic systems by opening a way towards collaborative planning along the entire network. Next to advances in communication and transport, computer-based decision-support systems have fundamentally changed the operations management of logistic processes. At their early beginning these systems were mainly used to minimize the transportation and inventory-holding costs. Based on the available information about detailed material flows, these systems can further support the order information management, the warehouse operation and the customer service administration today. Extending the focus of logistic decision support towards this direction eventually pursues the strategic objective of logistics management, namely to reduce the system-wide costs across the whole logistics network.

At their core decision-support systems have models and algorithms which carry the variables for planning, implementing and controlling the flow of goods. Often the variables are determined by simple heuristics which provide decisions based on rules of thumb. In advanced decision-support systems, tailored methods of Operations Research are applied to optimize the logistic processes. Unfortunately, such algorithms take advantage from aggregate data structures which are in danger to become useless by only slight modifications of goals or constraints noticed under a detailed view of material movements. Reactive decision making based on bounded anticipation is therefore hindered by the current state of methodological support.

In order to enable the shift from anticipatory logistics operation towards response-based logistics operation, robust algorithms capable of quickly processing detailed data in frequently changing environments are highly desirable. This work aims to widen the methodological support for managing supply chains into this direction. Our approach focuses on a class of techniques to simulate decision making processes which are based on external feedback. The first descriptions of such *adaptive systems* came from biology. In that context adaptation designates an evolution process which progressively modifies a structure in order to gain better performance. For more than two decades this principle has received tremendous attention in economics, computer sciences, engineering and other fields. Striking problem-solving capacities and versatility coupled with the convenience to simulate an evolution process have quickly spurred the application of so called *Evolutionary Algorithms* to a wide range of problems.

Evolutionary Algorithms essentially are adaptive systems. They are able to change the behavior of a system in respond to changes in their environment. While we can run them as long as we like, they offer a solution at any time. For this reason Evolutionary Algorithms can be efficiently used to support reactive problem-solving of actors in the logistics network. Following modern terminology, we refer to Evolutionary Algorithms as *adaptive or learning agents.* Different to other software artifacts also designated as agents in the current literature, our adaptive agents do not communicate directly by exchanging messages. They communicate by mutually changing details of the common environment they perceive. The availability of shared information enables this collaborative planning of material movements between adjacent actors in the supply chain.

The book is divided into two parts. Part I lays down the fundamentals of evolutionary adaptive search before an architectural model for adaptive agents is developed. Part II is devoted to applications of adaptive agents for planning and scheduling of logistic activities. These processes have often extremely difficult to solve optimization problems at their core. By focusing on elementary yet complex operations, such as grouping, routing, packing and scheduling, it is hoped to glean sufficient details of realistic problems.

Christian Bierwirth
Bremen, August 1999

Acknowledgments

This book is the outcome of my research done at the faculty of economics, University of Bremen, Germany, from 1993 to 1999. There are several people who accompanied the work over the years.

The holder of the chair of logistics at the University of Bremen, Herbert Kopfer, kindly invited me to discuss many of the ideas which are captured in the book. Hans-Dietrich Haasis (University of Bremen), Hermann Gehring (University of Hagen) and Richard Vahrenkamp (University of Kassel) undertake an assessment of the first version of the text. My former colleagues Klaus Schebesch, Thomas Utecht, Ivo Rixen and Annette Blome were involved with me in several research and application-oriented projects, which served as valuable input for some of the chapters in hand. However, to lay down the book took much more time than I ever had planned before. If it is amenable to a broader audience today, this is due to my friend and colleague Dirk Mattfeld who encouraged me to revise my schedule countless times. I would like to thank all people mentioned above very, very much.

I

FUNDAMENTALS OF EVOLUTIONARY ADAPTIVE SYSTEMS

Chapter 1

FROM ARTIFICIAL TO COMPUTATIONAL INTELLIGENCE

Traditional *Artificial Intelligence* (AI) claims the methodology of *rule-based systems* to be one of its leading programming paradigms[1]. Rule-based systems, also known as production systems, have been developed in the late sixties in order to provide a flexible representation of knowledge. They take advantage of deductive logic which is handled efficiently by symbolic data processing. Rule-based systems mainly became popular for the construction of Expert Systems which had a rapid spread at that time. In the seventies rule-based approaches run into their boundaries. The construction of decision making systems according to principles of human intelligence failed because of a lack of an adequate mechanism to extend available knowledge. A remedy was assumed from the incorporation of numerical knowledge representations. Some of the ideas developed at that time aimed at modeling processes of inference and learning on a computational basis. To stand out from the traditional symbolic AI, these approaches are captured by the modern term *Computational Intelligence* (CI).

Before we start reviewing different learning techniques of Computational Intelligence and their mutual relationships, we take a closer look at one of its roots. The chapter is closed by an overview of the book.

[1] cf. Nilsson (1980), Chap. 1.

1. BACKGROUND

From a philosophical point of view, learning is closely related with the justification problem of induction. We take up this discussion to work out the background of current research in Computational Intelligence.

1.1 RULE-BASED SYSTEMS

The *knowledge base* of a rule-based system simultaneously supports information and rules. Facts about a given environment, like for instance the data determining a certain situation, is designated as *information*. In difference, a *rule* represents domain knowledge about the environment under consideration. Rules are typically obtained through knowledge acquisition, so as to consult human experts or the like. Rules are given as *if-then* clauses consisting of a condition part and an action part. A rule is invoked if its condition part is satisfied either by information stored in the knowledge base or by facts detected in the environment. In the latter case new information about the environment is inserted into the system. Once a rule is invoked, the action encoded in the action part is carried out. Eventually, either information is added to the knowledge base or an action is effected in the environment.

Detecting facts, applying rules, and effecting actions describes a process of making a series of *inferences* in sequence. A so called *inference engine* is responsible for the control of this process. If all inferences are triggered by information the engine is called *data-driven*. Often several rules are invoked simultaneously and therefore inference engines involve a conflict solver. Typically, this component performs a goal-directed search to determine the precedence among the invoked rules. During the system's runtime the knowledge base expands. The inference engine is therefore often called a "learning component".

The convincing advantage of rule-based systems is the possibility to generate actions or decisions perfectly on the basis of all the knowledge available. If all rules passed to the knowledge base are accepted without restrictions, then failures or shortcomings of the system can be caused only by a lack of knowledge. The question whether rule-based systems are truly able to extend knowledge by using a data-driven inference engine remains a point of issue. Deciding on this question in the end means to grant rule-based systems the capability to learn.

The question raised is one of principal concern much more than simply to judge a paradigm of AI. In terms of logic an inference proceeds from one or more premises to a conclusion. That is, if we accept the premises we have to accept the conclusion. Notice that both types of knowledge to be found in rule-based systems, rules and information, can be regarded as premises of an inference. Consider the following argumentation.

Premise A1: *If a person wears a wedding ring then it is married* (rule).

Premise A2: *This person wears a wedding ring* (information).

Conclusion A: *This person is married.*

An inference of this kind is called a *deductive argument.* A deductive argument is valid if the conclusion inevitably follows from the premises and invalid if it does not. The validity of deductive arguments is discovered by matching the available information and the condition part of rules. Therefore data-driven inference engines can take advantage of symbolic data processing.

Right now we aware that all action effected by a rule-based system is established through deductive arguments. In contrast most of the inferences we make in common life are not valid deductive arguments. In common life we are interested rather in the correctness of a conclusion than in the validity of an argument. Suppose you want to know if someone is married or not. You may look for the wedding ring at a person's finger and conclude according to common sense that the person is married. Do we usually trust in this conclusion because we have made a valid deductive argument? Of course not. We trust in the conclusion because we usually accept the rule behind. The question raising now is why we do accept a rule, if it is not unlike to fail. The answer is simple. Our experience based on countless instances observed has tempted us to make an inference of the following kind.

Premise B1: *Peter wears a wedding ring and he is married.*

Premise B2: *Paul wears a wedding ring and he is married.*

Premise B3: *Marry wears a wedding ring and she is married.*

Premise B4: *I don't know anybody who wears a wedding ring and who is not married at the same time.*

Conclusion B: *All people who wear a wedding ring are married.*

Notice that conclusion B and the general rule A1 are equivalent. Nevertheless, the last inference is certainly not a valid deductive argument because the conclusion does by no means follow inevitably from the premises. Data-driven rule-based systems are therefore not able to derive this inference. By taking a closer look at the inference, we notice that it really extends knowledge because its conclusion goes beyond the premises. Actually the available information has implicitly invoked the following rule:

Premise B5: *If I know a large number of people who wear a wedding ring and who are married and I don't know anybody who wears a wedding ring and who is not married at the same time, then all people who wear a wedding ring are married.*

The conclusion above is drawn although the premises contain no quantification of the term "large number". Such inferences are called *generalizations* or *inductive arguments.* Inductive arguments offer the advantage to extend knowledge on danger to be seriously mistaken. In common life people use generalizations all the time. Even so it remains important to explain why and when circumstances make it reasonable for us to trust in the conclusion of an inductive argument.

This question was first raised in 1751 by the philosopher David Hume in his epochal work "*An enquiry concerning human understanding*". In modern epistemology the problem is called the *riddle of induction.* For short it can be stated as follows. On one hand there is a great need to use inductive arguments while, on the other hand, there are no logical fundamentals to do so. Unlike deductive arguments, inductive arguments need a justification. To justify an inductive argument means to give at least one good reason

> why a number of cases of a rule being fulfilled in the past afford evidence that it will be fulfilled in the future[2].

Hume categorically refused that any such reason could exist at all. In contradiction, modern philosophers predominantly claim a principal justification of inductive arguments[3].

An outstanding approach to solve the riddle of induction has been proposed by Goodman (1955). Different to other research in the field, Goodman did not primary attempt to justify inductive arguments. He rather showed a principle which underlies all kind of inferences, deductive and inductive ones. Goodman noticed that, in a way, also deductive arguments need a justification. The reason is that deductive arguments may be valid although their conclusion is wrong. In reverse, they may be invalid although their conclusion is right by accident. Therefore we have to reconcile the validity of any inference with the correctness of its conclusion by the following principle.

---

[2] from Chap. 6 "*On Induction*" in Russell (1912).

[3] For a collection of papers concerning this discussion see Swinburne (1974).

i. A rule is modified if it leads to a conclusion we are not willing to accept.

ii. A conclusion is refused if it contradicts a rule we are not willing to modify.

In result, justification of inferences is achieved by mutually adapting rules and conclusions. In other words, the way people make inferences, which we expect to be intelligent, can be interpreted as a continuous *process of adaptation.*

1.2 FUZZY LOGIC

Goodman's theory calls attention towards the necessity of drawing conclusions under uncertainty and towards the possibility to change rules. With these ideas in mind we return to our initial consideration of rule-based systems. In order to alleviate shortcomings of data-driven inference engines it appears reasonable to look for ways of representing uncertain knowledge and for ways to modify rules appropriately.

A straightforward approach addressed to represent uncertain knowledge is shown by probability theory. Uncertainty concerning a conclusion can be modeled by assessing data according to the relative frequencies of specific information. In our example it is for instance possible to determine the relative frequency of people who are known to be married, and the relative frequency of people who are known to wear a wedding ring. Moreover it can be expected that it is no problem to decide whether people who are known to be married wear a wedding ring or not. Hence the conclusion "*This person is married*", given the person in question wears a wedding ring, is assigned a probability which can be calculated by use of the Bayesian inversion law.

Nevertheless, there is a serious objection against the assumption that the correctness of a conclusion can be estimated appropriately on the basis of laws of probability. This is simply because uncertainty is often due to imprecision which is by no means of statistical nature. In general, uncertainty cannot be equated with randomness. Uncertainty which does not obey to laws of probability is commonly referred to as *fuzziness.* In mathematics fuzziness is treated within the theory of fuzzy sets which has been pioneered by the work of Bellman and Zadeh (1970).

Next to other applications the theory also had an impact on the development of rule-based systems[4]. Contrasting approaches based on probability theory, *fuzzy-logic* approaches permit to assign uncertainty directly

[4]For details of applying fuzzy logic to rule-based systems resulting in fuzzy-logic controllers and fuzzy Expert Systems see Zimmermann (1987).

to a rule. As already mentioned, rules given by *if-then* clauses are equivalent to statements quantifying over a set of objects like "*All people wearing a wedding ring are married*". The central idea of fuzzy logic is to define *fuzzy quantifiers* which can replace the ordinary quantifiers of deductive logic.

Fuzzy quantifiers are introduced by formal definitions of the weak quantifiers used in common languages, such as "few", "many", "seldom", or "frequently". A mathematical definition of these and other *linguistic variables* is obtained by a functional description for the grade of membership of an element $x \in X$ in a subset $A \subset X$. The *membership function* μ_A of set A ranges over a subset of the nonnegative real numbers. Often $\mu_A : X \to [0, 1]$ is used. If a membership function takes only values of either 0 or 1 for all $x \in X$ it is identical to the characteristic function of A over the universe X which is defined by ordinary set theory. Otherwise set A is referred to as a *fuzzy set* and the membership function μ_A can express the desired importance, weight, or urgency of a linguistic variable related to A. E.g. a membership function μ_A with $\mu_A(x) = 1$ $(x < 5)$, $\mu_A(x) = -0.2x + 2$ $(5 \leq x \leq 10)$ and $\mu_A(x) = 0$ $(x > 10)$ may define the quantifier "few" over the set X of natural numbers. Now, an inference from a *fuzzy rule* based on a fuzzy quantifier comes close to the usage of generalizations in common life. There, rules rather express rules of thumb like "*Almost all people wearing a wedding ring are married*" than definite assertions.

Contributions of probability theory and fuzzy logic to computer-based inference making can alleviate many weaknesses of conventional rule-based systems. The common principle aims to mimic the character of inductive arguments through inferences based upon vague interpretations of either information or rules. Technically this is implemented by adjusting numerical parameters of probability or membership. Decision making systems including such mechanisms often improve significantly.

1.3 CLASSIFIER SYSTEMS

The approaches described so far cannot completely satisfy because they do not explain how knowledge grows, even in face of uncertainty. Traditional AI has predominantly focussed on the formal aspects of inferences which only bear little relation to how people learn. In conclusion one can state that traditional AI has failed to create computer systems capable of improving themselves. In order to capture the ability of self-improvement for a system the following questions must be answered[5].

[5] cf. Holland et al. (1986).

- How can the system determine that rules are inadequate?
- How can it generate plausible new rules to replace the inadequate ones?
- How can it refine rules that are useful but non-optimal?

These questions bring us back to the justification problem of inferences. An inference, and in particular an inductive argument, is justified on the basis of an adequate rule. But how can a system decide whether a rule is adequate? Nowadays we do not believe that these questions can be answered by a formal treatise of inference processes. To justify a generalization actually means to accept the common practice of its use. If we follow this claim another question remains open.

- Does there exists a general procedure for justifying inferences which is so rigorous, that it can be mapped into a computer program?

Indeed there is. This answer given by John Holland (1975, 1992) is discussed in greater detail later on, thus only a few words are said here.

Holland constructed a system he named *Classifier System*[6]. A Classifier System (CS) is a conventional rule-based system with two further components added. One component, called the *Bucket Brigade Algorithm*, determines the worthiness of rules by representing their adequacy. Holland assumed that the worthiness of a rule changes as a function of the frequency it is used. The more often a certain rule is invoked, the higher its worthiness is expected. The other component functions for the discovery of new rules. Holland named it *Genetic Algorithm* because it enables to generate new rules from the rules given in the knowledge base by applying operators which model biological reproduction processes. The frequency the Genetic Algorithm selects a certain rule for the generation of new rules is controlled by its observed worthiness. With regard to the central role of *selection* in nature, Holland refers to the worthiness of rules as their *fitness*.

1.4 ADAPTATION AND LEARNING

It is interesting to aware that Classifier Systems agree with the theory developed by Goodman (1955). A CS realizes the principle to reconcile knowledge through its use. We have already recognized this as a continuous process of adaptation. Today, adaptation is widely accepted as a driving force underlying processes of search and learning in nature. This

[6] The name refers to the desired capability of rules to assign information to arbitrary attribute classes such as "*married*".

understanding, however, does not explain how learning actually can take place. In order to get an idea we may ask how learning is explained in cognitive sciences.

Learning, the extension of knowledge, is generally assumed to base on feedback regarding the success or the failure of a certain behavior or concept. Learning processes are controlled by the individual abilities to access such concepts, which is referred to as *problem-solving.* Inductive learning, for instance, is seen in the derivation of general concepts from well known examples which have to prove themselves against experience.

In computer systems adaptation is approached by the adjustment of parameters. If we want to simulate learning we depend on numerical data processing in same manner. This explains the rise of the term CI in recent literature. In allusion of the dominant role of symbolic data processing in traditional AI, the term CI points out the change of the paradigm. According to Eberhart (1995), *Computational Intelligence* is defined as

> a methodology involving computing that exhibits an ability to learn and/or to deal with new situations, such that the system is perceived to possess one or more attributes of reason, such as generalization, discovery, association and abstraction.

2. LEARNING BY COMPUTATIONAL INTELLIGENCE

The most widespread fields of CI are *neural computation*, *evolutionary computation*, and *fuzzy logic.* In order to facilitate a comprehensive understanding of the different learning techniques we start with a brief description of neural computation and evolutionary computation. These fields are devoted to the design and application of *Artificial Neural Networks* and *Evolutionary Algorithms.* In Sect. 1.2 we gave already a brief outline on fuzzy logic for application to computer-based inference making. Applications of fuzzy logic combining other techniques of CI are briefly reviewed in this section.

2.1 NEURAL COMPUTATION

An Artificial Neural Network (ANN) is a system which roughly models the massively parallel structure of the human brain[7]. The methodology basically serves for an approximation of computational functions mapping an argument space into an image space. Thus one can make use

[7]For a comprehensive introduction to neural computation and its main fields of application see Hertz et al. (1991).

of neural computation in solving related tasks like forecasting, pattern recognition, and association.

The *architecture of an ANN* is defined by a directed and typically highly interconnected graph. The input-output interface of an ANN is represented by two sets of nodes in the graph, the set of source nodes and the set of sink nodes. These sets are respectively referred to as *input layer* and *output layer*. All other nodes are located at so called *hidden layers*. If the graph contains no cycle the ANN is called *feed-forward* and *recurrent* otherwise.

The nodes of the graph, also referred to as *neurons*, represent the processing elements of the ANN. Each neuron contains a vector of *connection weights* which expresses the *synaptic intensities* of its connecting links coming from other neurons. Internally a neuron intensifies an input signal by multiplication with the weight of the corresponding link. In order to determine the activation level of a neuron all its weighted input signals are summed up. Finally an *activation function* transforms the activation level into an output signal. The activation function often includes a threshold, i.e. neurons cannot become active until a certain level of input intensity is reached.

Any ANN is basically defined by its architecture, a set of weight vectors and an activation function. From a mathematical point of view an ANN represents a transformation of a set of input signals into a set of output signals. From theory it is known that an ANN can approximate in principle every transformation function with arbitrary precision, given the number of hidden neurons is large enough. Unfortunately this does not imply that any number of neurons or even any particular architecture is designated to approximate a given function properly. In practice neural computation is particularly useful for the approximation of non-linear functions which are substantially unknown, although a lot of example data is available[8].

Suppose there is a particular ANN with a certain architecture and an appropriate threshold activation function given. The variable parameters of the system are the weights of all links to be found in the net. Suppose further that the ANN shall approximate a given time-series of a finite number of pairs $(t, x_t), 0 \leq t \leq n$. Thus the weights of the links in the ANN have to be chosen such that an input value of t leads to the desired output value x_t. This is achieved by a process of training the net which is called *supervised learning*. First the input vector (t) is fed into

[8] ANNs have been successfully applied to diverse quantitative problems in management science. They are of special interest in forecasting of financial time-series, see e.g. Rehkugler and Zimmermann (1994).

the net and the output vector ($\tilde{x}_t$) is taken. Then the output is compared with the desired output resulting in the error vector ($x_t - \tilde{x}_t$). Finally the observed errors are *back-propagated* through the net. The synaptic intensities of neurons are readjusted such that the deviation from the desired output is minimized. Different gradient methods reducing the error-rate can be used, dependent on whether the net is recurrent or not. Following the phase of learning, one then expects the ANN to predict output signals beyond the example set on a reliable basis.

Contrasting to tasks of forecasting, other types of problems do not contain explicit example data. Assume an ANN is used to solve e.g. an optimization problem. Here, one cannot make use of supervised learning algorithms anymore. Here, other principles of reinforcement learning have been adopted to fit the *connectionist* models of ANNs as well. However, the quality of recent neural-computation approaches in optimization is predominantly weak.

It is beyond the scope of the book to present the diverse approaches in detail. But it is important to mention that the structure of the learning process involved always remains the same. In the terminology of neural computation "learning" is used as a metaphor for adjusting a multitude of numerical parameters in parallel so that a desired computation is performed. In this way a specific problem can be solved by a network connecting a lot of simple processing elements. Although this idea was realized already in the early sixties AI hardly showed interest in neural computation for a long time. The possibility to learn environments through connectionist models was underestimated until efficient algorithms for error back-propagation became popular[9].

2.2 EVOLUTIONARY COMPUTATION

Apparently different from neural computation, another field of CI models search processes which are known from natural evolution. The most striking characteristic of the techniques establishing Evolutionary Algorithms (EAs) is that they work on a *population* of search trials, rather than on a single state. Members of the population, so called *individuals*, encode a feasible input of the surrounding environment. This input may be for instance

- a candidate solution of an optimization problem,
- a decision or policy of a control problem,
- a strategy of an agent in an economic game,

[9] cf. Rummelhart et al. (1986).

- domain knowledge gathered by a rule-based system.

The suitability of every input is tested under the circumstances found in the environment. On this basis a *fitness value* is assigned to the individuals of the population. Starting with a randomly generated population, new individuals are generated through modifications of existing individuals. The higher the fitness of an individual is the more often it is selected for the modification process. This iteration is carried out until a prescribed stopping criterion, e.g. a fixed number of *generations*, is reached. After a number of generations carried out one hopes to find well adapted inputs of the environment in the population.

In Sect. 1.3 we have mentioned Genetic Algorithms (GAs) which are used for the discovery of knowledge in a Classifier System. The discovery of this kind of algorithms can be regarded as a side effect of the effort made to improve rule-based systems. But it was also recognized by Holland (1975, 1992) that the method can be applied directly to a much wider range of problems. GAs actually have been proved to be applicable in all of the problem fields listed above although there is a clear focus on optimization[10].

Often GAs are even equated with function optimizers because the idea of evolutionary search has a convincing interpretation for optimum seeking. Suppose a function to be maximized. The fitness of an individual may express the value of this function at the argument encoded by the individual. In this way individuals represent candidate solutions for the problem. Since we start to run the algorithm with random solutions, optimization is depicted as a search process which is guided by the quality of solutions met.

There is no definite classification scheme for EAs at the present time. Traditionally EAs are distinguished by the different application fields originally addressed. There are a few but substantial differences concerning the way how individuals are represented and modified. The following overview briefly outlines the prevailing branches of evolutionary computation[11].

Genetic Algorithms are predominantly used for constrained optimization problems. In the standard GA individuals are represented by fixed-length strings over a finite alphabet. Most widely used is a binary alphabet with the exception of a CS where the GA uses a ternary alphabet. A low-level representation is necessary because new strings

[10] The application of GAs to optimization problems has a vast and confusing literature. For a survey the reader is referred to the regularly updated report series of Alander (1998).
[11] For more details see the survey of Bäck et al. (1997).

are assembled from pieces taken from strings already existing in the population. In reference to its biological counterpart this operation is called *crossover*. Two GA variants are of further practical relevance.

Order-based Genetic Algorithms extend the GA approach into the space of permutations. In this way combinatorial order problems are made accessible without mapping these problems into a pseudo-boolean representation.

Genetic Programming (GP), like Classifier Systems before, is addressed to application in machine learning. Different to a CS, GP does not evolve rule-based domain knowledge but computer programs, capable to solve a given task. The idea is due to Koza (1992) who showed that certain tree structures, suitable to modifications by crossover, can be interpreted as expressions of a programming language. Most widely used are Lisp code representations.

Evolution Strategies (ES), as introduced by Rechenberg (1973) and extended by Schwefel (1977), Schwefel (1995), have been developed independently from GAs at nearly the same time. ES were originally designed to control experiment-based parameter optimization in engineering applications. Meanwhile they are proved to be powerful in computer-based optimization as well. An ES modifies the individuals by means of a so called *mutation* operator. A mutation changes an individual to a certain extent by control of a step-size mechanism. Since the individuals of the population are mutated independently from each other, a high-level representation of individuals by real-valued numbers is suitable.

Evolutionary Programming (EP), introduced by Fogel et al. (1966), is the oldest EA representative, but initial it received little general attention. EP was created to evolve prediction systems based on finite-state machines. Like other potential applications of EAs, this problem cannot be mapped directly to one of the above mentioned representations. In order to evolve for instance variable structures like ANN architectures, non-standard representations are proposed. Using a problem-specific representation, one typically cannot involve a crossover operator by avoiding infeasibility of individuals at the same time. A problem-specific mutation operator is used instead which effects feasible changes of individuals at random. Evolutionary computation dealing with the optimization of variable structures shows apparent similarities to Fogel's approach. Therefore it is commonly referred to as EP.

According to Bäck et al. (1997), three main-streams are distinguished in evolutionary computation, namely GAs, ES, and EP. From theses types diverse variants can be found in recent literature addressing computer science, engineering, natural science, economics and management science. Differences are observed mainly in the meaning and representation of individuals, in the way the modification operators work, and in the mechanisms responsible for the selection of individuals that undergo a modification.

For the moment we neither discuss design decisions necessary to derive competitive algorithms from evolutionary computation nor emphasize particular application fields. This is because we think that every successful EA is an instantiation of a framework underlying evolutionary computation in general. The framework emphasizes adaptation by means of selection. Once such a framework has been developed, it can be incrementally instantiated in order to fit the needs of a particular application problem.

2.3 COMBINED APPROACHES

In order to approach the capabilities of naturally intelligent systems it appears attractive to look for a methodological synthesis of CI. Although one may doubt that the framework assumed for evolutionary computation can be extended to capture neural computation and fuzzy logic, there exist important relationships between these methods[12]. Most of all, the common use of numerical knowledge representations is seen as a basis to combine the actions of different CI methods. In the following we sketch the prevailing approaches to connect EAs with the other paradigms.

NEURAL AND EVOLUTIONARY COMPUTATION

A prerequisite for using ANNs is the construction of a network architecture which fits the requirements of a specific task. If there are too many connection weights which have to be adjusted, the error-rate typically converges very fast. In case that the example data is somehow noisy the net tends to learn the noise then. Consequently an ANN looses the capability to make extrapolations of the example data on a proper

[12]The matrix model of Bezdek classifies relationships among system components by two attributes. One attribute refers to the carrier of knowledge, which is either numeric, symbolic, or organic. The complexity increases from the numeric to the organic level. The other attribute addresses the systems purpose, representing neural processing, pattern recognition, or intelligent behavior with increasing complexity as well. For a discussion see Eberhart (1995).

basis. In reverse, too few connection weights prevent that the error-rate drops below an acceptable level. In turn this leads to prohibitive long and disappointing phases of learning.

In order to discover efficient architectures for ANNs several EAs are proposed in literature. An approach to design feed-forward structures by GAs is given by Harp et al. (1989). A binary representation of net structures, suitable to modification by crossover, encodes the number of layers, the number of neurons per layer and the connectivity from one layer to the other. In the approach of Kitano (1990) a recurrent network structure is optimized by means of EP. This is possible by using a specification of architectures in terms of a graph representation grammar.

Typically such approaches use supervised learning by error back-propagation for the adjustment of the weights in a network architecture. After this training the fitness of an architecture is determined by the performance of the ANN, approximating a subset of the example data not used before. As already mentioned, an important application domain of ANNs is seen in predicting time series. Currently, approximation models evolved by EAs belong to the best forecasting methods available[13].

In reverse, evolutionary computation can also interact with ANNs to find suitable connection weights. In this case a fixed net architecture is considered. The EA functions as a learning algorithm for the reduction of the error-rate of the ANN. Caudell and Dolan (1989) apply this idea for supervised learning. In their approach a GA searches the multidimensional real-valued space of potential connection weights. The performance of the ANN is measured in terms of the mean-square error observed for the given set of example data. This value determines the fitness of the specific weight vector used. In an approach of Whitley et al. (1991) the same idea is applied to an ANN incorporating reinforcement learning.

Recent approaches to use evolutionary computation for the construction of learning algorithms have led to moderate results. EAs are mainly used as a base technique to optimize the performance of a particular ANN. Such approaches are capable to solve only small problems on a reliable basis, like to recognize simple binary patterns or to approximate the exclusive-or operator. Turning to larger problems, ANNs which use gradient-based learning algorithms usually produce superior results. Nevertheless, connecting neural- and evolutionary computation for machine learning is still an active field of current research.

[13] cf. Harrald and Kamstra (1997).

FUZZY LOGIC AND EVOLUTIONARY COMPUTATION

Like in traditional Expert Systems, the knowledge base of a fuzzy-logic controller can be obtained by an analysis of expert experience. As we have mentioned in Sect. 1.2, a fuzzy system is based on fuzzy rules which express certain control mechanisms on a linguistic basis. Facts detected in the environment of the system are brought into contact with theses rules by validation of membership functions. An interesting alternative appearing when experts cannot do it otherwise is to evolve a fuzzy system by means of an EA. Similar to interactions of EAs and ANNs, evolutionary computation can contribute to the field by learning suitable fuzzy rules containing linguistic variables or by adjusting numerical parameters in the membership functions involved.

A general approach to learn a set of fuzzy rules has been proposed by Thrift (1991). Suppose there is e.g. a two-dimensional control problem given. Suitable rules of the form *"if Parameter 1 is X and Parameter 2 is Y then turn the control parameter to Z"* are searched, where X, Y, and Z refer to particular linguistic variables. If the parameters come from linearly ordered spaces the corresponding linguistic variable can be described by certain attributes like "positive, negative" or "large, medium, small". In a next step the combinations of attributes, e.g. "*negative and large*", are expressed as fuzzy sets by definition of fixed membership functions. Thus the entire range of a parameter is mapped into a finite chain of overlapping fuzzy sets. In this way each of the linguistic variables X, Y, and Z are modeled. Then a two-dimensional matrix over the sets of attribute combinations defined for X and Y is built up. The elements of this matrix are from the set of attribute combinations defined for Z, i.e. each entry in the matrix stands for a triplet of attribute combinations. Since each such triplet is associated with three particular membership functions, a triplet can represent a fuzzy rule at the same time. Thus any matrix constitutes a complete two-dimensional fuzzy decision table. Finally a GA is used for searching the space of matrices in order to learn a suitable rule-base for a fuzzy-logic controller.

In reverse to the approach of Thrift who uses fixed membership functions, Karr (1991) proposes evolutionary computation for learning membership functions while leaving the fuzzy rules fixed at the same time. Linguistic variables modeled for fuzzy-logic controllers are often defined by membership functions shaping a simple triangle over the parameter space. Eventually a membership function is determined by the anchor points locating a corresponding triangle. Since its coordinates are real-valued numbers they can be directly used to represent candidate membership functions. However, Karr decodes a membership function once more by mapping the coordinates into a fixed-length string over a binary alphabet. Then a GA is applied in order to optimize the performance of a fuzzy-logic controller.

Connecting both approaches, EAs can be used to learn fuzzy rules and their accompanying membership functions simultaneously. But empirical tests show that integrating the design stages of a controller is a difficult task for adaptation. Due to the complex interactions of rules and membership functions, the size of the search space increases drastically, if a complete coverage of the rule base is intended by using fuzzy-decision tables. A remedy, proposed by Lee and Takagi (1993), bases on the observation that large parts of fuzzy-decision tables often contain superfluous rules. In order to reduce the number of rules to a few and hopefully suitable ones they use a variable-size representation of the rule base. Consequently evolutionary computation deals with a variable structure and EP is the method of choice.

Approaches of the above kind are mainly tested by generating fuzzy-rule bases for the control of technical systems. To hold for instance a mechanical system in an state of balance under disturbance is an extremely difficult problem in general. Here, control systems are of high practical relevance and much effort is spent in engineering to develop new design methods. By taking advantage of evolution-based design methods the performance of a fuzzy-logic controller can improve strongly[14].

3. OVERVIEW OF THE BOOK

Throughout this chapter we have introduced CI as an approach providing different techniques for learning in artificial systems. Many of the fundamental ideas have been developed already thirty years ago, but their high computational demand hindered a thorough validation and immediate acceptance.

Today, management science is one of the most profiting target fields of CI. The application of intelligent methods for the decision support of industrial enterprises has been recognized as a business necessity. The management of financial transactions, work-flow organization and distributed production processes require disaggregated forecasting, planning and control structures. Improving the efficiency of these business operations promises important competitive advantage. However, anticipating and effecting detailed states of the underlying logistics systems leads to quantitative problems which are often poorly structured, intractably hard or prohibitively time consuming to be solved by tradi-

[14] For a detailed treatise of techniques connecting fuzzy logic with evolutionary computation and neural computation, see Jang et al. (1997).

tional methods. The remedy promised by robust and versatile CI methods therefore appears highly attractive. Unfortunately the current state of CI technology lacks a profound theory about its prospects and limits. This means that we have to choose a method only on the basis of previous computational experience. The vast majority of experimental studies in the field in turn indicate a domain-dependent suitability of the available techniques.

- Fuzzy approaches have improved the control of technical systems.
- Neural computation has caused advances in forecasting time series.
- Evolutionary computation has been successfully applied to many hard optimization problems.

Logistics management aims at optimizing the customer-order process by planning, implementing and controlling an efficient flow of goods. Consequently we concentrate on the evolutionary branch of CI technology in the following. The book is organized in two parts. Part I lays down the fundamentals of evolutionary adaptive search while later on Part II deals with the incorporation of these methods into logistics management systems.

Chap. 2 analyzes the structure of adaptive systems and discloses their interfaces. The resulting framework enables the definition of independently acting entities which solve problems cooperatively under a certain coordination of action. Examples of distributed logistic planning are outlined.

A brief introduction to evolutionary computation is given in Chap. 3. The concept of adaptive search is pointed out by the general idea of GAs. Features being of particular interest like problem representation issues, constraint handling and the incorporation of domain knowledge are presented.

In Chap. 4 we focus on the structure of combinatorial search spaces which frequently underly logistic problems. Analyzing the search space for a problem at hand helps predicting the success of adaptive search. A positive assessment is demonstrated for the famous Traveling Salesman Problem.

The capability of evolutionary algorithms to cope with dynamically changing environments is subject to Chap. 5. This perspective introduces the idea of reactive agent behavior. For an example of collaborating agents we consider a complex problem of production planning which integrates the lot-sizing and the scheduling level. A framework for adaptive agents is finally depicted which accompanies us throughout the second part of the book

The creation of adaptive agents requires a formal representation of their field of application. The first chapter of Part II (Chap. 6) analyzes the representation structure of logistic base operations such as grouping, packing, sequencing and scheduling of goods or activities. Then we provide a common representation scheme for the corresponding optimization problems which enables the definition of similar yet problem-effective search operators.

This approach is tested in Chap. 7 for one of the considered problem fields in greater detail. Using a standard model of job-shop production, we analyze search spaces of scheduling problems before we compare the performance of adaptive search with the outcome of advanced problem-specific methods. Although these tailored methods outperform our approach slightly, the proposed GA turns out robust even if complicating realistic features like product delivery-dates or customer priorities are taken into consideration.

In Chap. 8 we show how a simple domain-specific heuristic can be engaged to improve the adaptive scheduling algorithm considerably. By tuning the algorithm towards a good compromise of runtime and solution quality, the approach effects highly efficient processes of scheduling and rescheduling.

The suggested algorithm is used in the last chapter as a learning method inside of an adaptive agent. Perceiving a changing production environment, the agent has to respond adequately to unforeseen events like incoming orders and downtime of resources. In a case study from industry we investigate its capabilities to reactively plan the day-to-day activities of a workcenter. Finally, the approach is applied to an on-line problem in automated manufacturing. To accelerate the reaction times of the agent, state changes of the environment are monitored by its integral memory component. A superior control is achieved this way which dominates the best-known conventional control methods.

Chapter 2

PRINCIPLES OF SYSTEMS

The feedback mechanism involved in all principles for artificial learning refers back to the notion of the control-circuit. This very common concept serves as a basis for the analysis and the design of feedback-control systems.

The foundations of systems are briefly reviewed in the first section. In the second section we concentrate on the decomposition of complex systems into coupled subsystems. From a structural view, a complex system consists of a number of components which influence each other by means of mutual interventions. The overall system control is typically decentralized and therefore it requires a coordination in order to work reliably. We approach a coordination system, capable to capture different concepts of coordination within a common hierarchical framework. In the third section we focus on decision making systems. Three planning problems related to the manufacture, storage and distribution of products are given. It turns out that different strategies are needed to obtain a suitable coordination of action for each of the complex logistic tasks. This leads us to three principles of coordination which are formulated upon the framework introduced before. Finally, the approach is compared with the common classification of decision making systems.

1. SYSTEM STRUCTURES

A structural classification of systems is presented. Starting with a definition of general systems the feedback-control principle leads us to a base model of adaptive systems. Special emphasize is put on internal models, a central aspect when reflecting computational learning later on. Finally, two fundamental ways of external interventions are identified for adaptive systems.

1.1 GENERAL SYSTEMS

The term *system* is common in many contexts but even in science it is differently used. In order to contain a segmentation and to provide a classification of systems the *general system theory* has been developed[1]. In great generality a *system* is described as a set of *system components* which are thought of as a whole and which are delimited from an exogenous environment[2]. The components of a system are interrelated in order to perform a certain task.

A formal definition of systems serves as a starting point of this structural approach. The *general system* S is a set of relations on a family of nonempty attribute sets V_i, $i \in I$

$$S \subseteq V = \times\{V_i \mid i \in I\} \qquad (2.1)$$

where "×" denotes the Cartesian product and $I \neq \emptyset$ an index set. The elements of V_i represent possible attribute values for the i-th component of the system. The *behavior* of a system is determined by the set S, i.e. by the possible combinations of attribute values of all components. The *structure* R^S of S denotes the set of existing pairwise relationships between any of two system components, i.e. $R^S \subseteq \{(V_i, V_j) \mid j, i \in I, i \neq j\}$.

The approach allows us to classify systems by their structure, i.e. by considering the relationships between components while neglecting the properties of these components. Of practical interest are *input-output systems* with a partitioned index set $I = I_x \cup I_y$, $I_x \cap I_y = \emptyset$ such that $S \subseteq X \times Y$ holds. Eventually S is a subset of the Cartesian product of just two attribute sets containing the various input and output values. Moreover, if S is a mapping of input arguments from X into the set Y of output values, the system

$$S : X \to Y \qquad (2.2)$$

[1] cf. Mesarović and Takahara (1975).
[2] cf. the German standard DIN 19226.

is referred to as a *functional system.* Notice that input-output systems totally hide the inside view on system components. For this reason they are also called black-box systems.

Although many other classes of systems can be distinguished, the scope of questions which are treated properly on the level of general system theory is very small[3]. This situation changes if we pay particular attention to systems containing *feedback-control* structures. This subclass of general systems is often seen as the true starting point for system unification.

Different to set-theoretic criteria, systems are also classified according to their purpose, i.e. by the kind of task they perform. On one hand a task can be given as a series of definite instructions. In this case we face an input-output system because the task consists of a well-defined transformation of inputs into outputs. On the other hand a task can be present in reference to a particular goal. Then the system performs a goal-seeking process and the objective is to reach the given goal as close as possible. In general, goal-seeking is regarded as a search process within a state space. Therefore goal seeking is often equated with problem solving.

Let us have a closer look at the behavior of a system performing a goal-seeking task. Typically goal seeking involves two major activities. First a search trial has to be generated. Then a response regarding the success of the trial is evaluated. Thus we face two central components inside a goal-seeking system: an active one, responsible for generating and evaluating trials and a passive one, representing the object system. Both components are interrelated by an internal loop of operation.

The notion of goal seeking leads us to the well known *control-circuit* principle. It represents the general feedback-control structure found in *feedback-control systems*[4]. The active component of a goal-seeking system is called *control system* or simply *controller.* The passive object system (being the object of control) is referred to as *process.* It is assumed to return a *feedback* regarding the success of control in the goal-seeking process. This interrelation between a control system and its process is shown in Fig. 2.1.

Let us consider the control system block. It receives an external *reference input* representing a goal to be reached. As long as this objective keeps unchanged the input is assumed constant. The input serves as a

---

[3] A survey on system types and properties derived on a set-theoretic basis is given by Guntram (1985).

[4] Due to the central role of control in cybernetics and engineering feedback-control systems have a rich literature, see e.g. van de Vegte (1990) for more details.

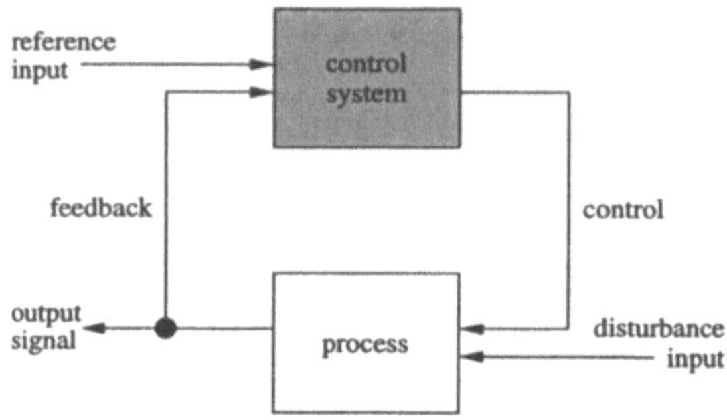

Figure 2.1 The general control-circuit structure.

reference of comparison to the controller. The deviation between the reference input and the actual state of the object system is transformed into a *control* which causes a certain manipulation of the process. The controlled process is thought of as being embedded in an external environment. Therefore *disturbance input*, which is due to changes in the environment, can effect the controlled behavior of the process. Consequently the actual state of the process can differ from the steered reference. The *output signal* produced by the process represents its actual state. This signal is either directly returned to the controller, or, as indicated by the black circle, it serves as an argument for an additional functional subsystem producing the *process feedback*.

Example 2.1 Let us consider a retailer who maintains the inventories of particular products. Acting as controller the retailer's objective is assumed in keeping certain safety stocks at any time. Consequently, the object system is represented by the inventory levels of the products. The control task is to decide on the volumes of supplementary orders which periodically replenish the stocks. The retailer has only information concerning the probabilistic distribution of demand, his decisions are based on expectations concerning the future demand. However, expectations can fail and some stocks run empty or full because of unforeseen or missing customer demand. These deviations of costumer demand correspond to disturbances of the object system here. The retailer receives a process feedback in terms of the deviations between actual and desired inventory levels. In reaction, the retailer possibly will change his expectations and modify the replenish policies. ∎

The example demonstrates that the behavior of feedback-control systems can be viewed at dynamically. At any time t the controller acts on the basis of its objective given and some internal parameters p_t, derived from earlier experiences. The control action effects an immediate manipulation of the object system. In consequence, the controller receives a feedback regarding manipulations within a short time lack Δt. Finally, the control parameters are adjusted with respect to the process feedback leading to $p_{t+\Delta t}$.

1.2 ADAPTIVE SYSTEMS

A system is called *closed* if there are no other relationships across the system's borderline to the environment with the exception of *disturbance*. In reverse a system is called *open* if it maintains relationships to exogenous objects in the environment. Feedback-control systems can be viewed as both, an open as well as a closed system. Like in the example above, control can be subject to disposition of a human operator. Due to various exogenous factors which simultaneously influence the operator's expectations, such systems are called open with respect to the definition. If the control action is generated automatically, the system is called closed in contrast. In this case a feedback-control system is referred to as an *adaptive system.*

In an adaptive system the process under consideration is the only component which is coupled with the environment. Basically, *automatic control* refers to internal control parameters which are automatically adjusted with respect to the reference input and the process feedback received. Thus, from conceptual point of view, an adaptive system maintains an *internal model* of the object system in order to adapt to the environment[5]. A human controller can build up a model of the object system as well, but the concept behind is different because such a model is by no means internal, it is part of the environment itself. In contrast, automatic control is viewed as a further closed-loop inside the process control unit.

The control system depicted in Fig. 2.1 is shown inside of the grey shaded box of Fig. 2.2. Internally, this system consists of two components, representing the parameter adjustment (active) and the internal model of the underlying process (passive). In order to adapt the internal model, a set of control parameters is adjusted in the inner loop of operation. Thereby a former state of the process serves as a reference to determine a feedback for each tried parameter setting. Control and feedback of the inner structure are called *ex-ante* because they have not effected the process thus far.

The sketched procedure simulates a process control by using an internal model. Whenever a parameter setting matches the given objective

[5]As outlined in Chap. 1, parameter adjustment is referred to as *adaptation* in computer science. The term adaptation actually stems from biology. Here, in a similar way, it designates a natural process of an organism fitting itself to its environment, cf. Maynard Smith (1989).

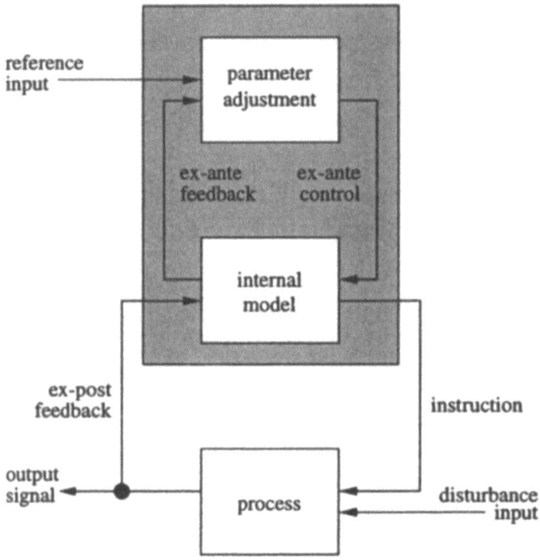

Figure 2.2 The general structure of adaptive systems.

(i.e. the internal model has been successfully adapted), an *instruction* is derived from the model and implemented in the process. Like before, the effect of control is returned but the feedback is *ex-post* this time. This feedback is necessary to reorganize the internal model and to refresh the process reference state.

There is a practical need to decide whether an adaptive system operates in actual time or not. In the latter case the adaptation of the internal model represents an operation to be performed not necessarily simultaneously with the generation of control instructions. Here the operation of components can be a stepwise procedure and, therefore, the adaptive system is called *off-line system.* In difference, a system operating in actual time is called *on-line system* and the operation of its components is regarded as a dynamic process.

To summarize, adaptive systems belong to the class of feedback-control systems. In order to generate control instructions automatically within a changing environment, adaptive systems maintain an internal model of the process to be controlled. The structure of adaptive systems is closed. Moreover the structure consists of two coupled control-circuits of which one adapts the internal model while the other one generates control instructions for the object system under consideration.

1.3 SYSTEM INTERVENTIONS

Due to the closed-loop structure of adaptive systems there are only two ways to access their components. One way is to influence the parameter adjustment component through a change of the reference input. The other way is to manipulate the internal model of the adaptive system by means of apparent disturbance. In system theory an intervention of the former kind is called *goal modification*, while an intervention of the latter kind is called *image modification*. To point out this difference we concentrate on the role of the intervening system by viewing the adaptive system as an object system.

GOAL MODIFICATION

A system intervenes an adaptive system by goal modification if it governs the reference input of the adaptive system. This situation is shown in the left diagram of Fig. 2.3. Both, the intervening and the adaptive system have their own reference input. But, as indicated by the bold arc, the intervening system can act as a controller and deal with the parameter adjustment component just like a process. In this context, parameter adjustment is sometimes referred to as an *auxiliary process*. The intervening system receives an ex-ante feedback (actually meant for the parameter adjustment component, as indicated by the dashed arc) after each intervention. It can be seen clearly that the adaptive control system is regarded as to act hierarchically subordinate. The goal modification principle can be interpreted differently:

i. The intervening system changes the reference regarding the adjustment of control parameters.

ii. The intervening system modifies the control-parameter space.

Case (i) is of particular relevance in organization theory where system interventions often express *principal-agent* relationships between individual components[6]. The principal (having the role of the intervening system) derives a number of subgoals from its overall objective at first. Then the pursuit of these goals is delegated, one after the other, to an agent (represented by the adaptive system). Hence interventions are carried out to initiate and coordinate the activities of organization members.

Case (ii) is of practical importance for the design of technical systems. Engineers use hierarchically coupled controllers in order to achieve a fine-tuned parameter adjustment: the superordinate controller manipulates the control-parameter space of the subordinate controller which generates the actual process control again.

[6] In organization theory the goal modification principle is also going by the term *management-by-objective* cf. Poensgen (1980).

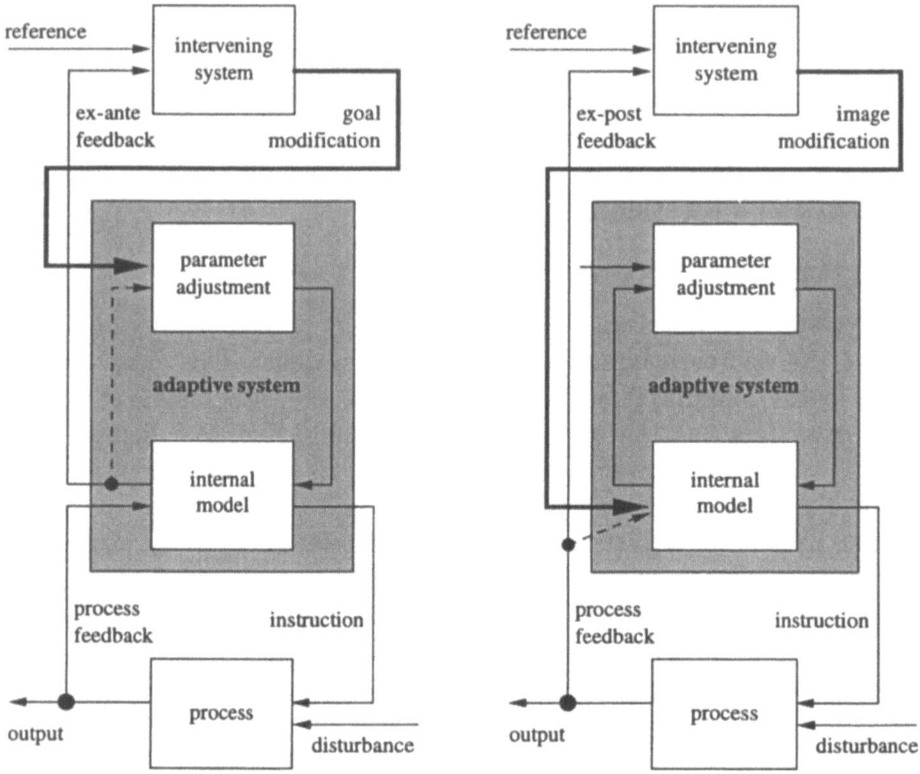

Figure 2.3. Adaptive system interventions: goal modification vs. image modification.

IMAGE MODIFICATION

A system intervenes an adaptive system by image modification if it can access its internal model. Unlike goal modification, image modification does not change the reference input of subordinate control. This situation is shown in the right diagram of Fig. 2.3. Recall that the internal model images a parameterized control function for the process considered. Within the inner loop of operation certain parameters are readjusted in order to adapt the internal model towards the needs of appropriate control. To accelerate adaptation it can be useful to change the internal model by circumventing the parameter adjustment component. Notice that an intervention by image modification effects the internal model in the same way like process feedback. Consequently, after each intervention by image modification, this ex-post feedback (meant for the internal model) is responded to the intervening system.

The image modification principle is interpreted in organization theory as well. There it stands for an exceptional intervention of a supervisor who temporarily takes over the function of subordinate organization members (*management-by-exception*).

Example 2.2 Intervention by image modification is comparable to a driving instructor who takes over control in a dangerous traffic situation. In contrast, to refer the learner just verbally at the risk of an accident represents an intervention by goal modification. ∎

Thus far we have outlined the structure of adaptive system interventions. Both types, goal modification and image modification, disclose a kind of hierarchical relationship between certain system components.

2. COMPLEX SYSTEMS

Complex systems consist of diverse subsystems which influence each other by mutual interventions. Therefore the analysis as well as the design of a complex system is a difficult task in general. Both, system analysis and system design basically aim on understanding what makes a system reliable in pursuing its objectives. The central topic in both fields addresses the question of problem decomposition and the coordination of decentralized system control. In order to approach complex systems we concentrate on these key features at first. Then we propose a conceptual framework, serving as basis for the design of complex adaptive systems.

2.1 DECOMPOSITION AND DECENTRALIZATION

Feedback-control systems are often so complex that they evade a complete understanding. This applies for instance to almost all natural systems. The difficulty in general is to differentiate between the simplicity of a system's description and the need to take the variety of its behavior into account.

A remedy of this dilemma is given by the *decomposition* approach. In order to reduce the complexity of a monolithic system its structure is split up into smaller but mutually dependent elements. Typically decomposition leads to a family of internal models[7], each concerned with a different aspect of the compound system. Then a system is said to be *decentralized* because it includes several distinct subsystems interacting in order to accomplish the common objective. Hence one can argue that each decentralized subsystem reflects a certain specialization of the overall system behavior.

[7]Like before, internal models are determined by a set of relevant parameters and transformation principles.

Example 2.3 Decentralized systems are met in every day life. They are most obvious in a business firm. Analyzing such structures we obtain organizations resulting from isolating specific jobs and assigning them to specialized members of the firm. In this way the firm's objective is broken down into a number of subgoals. To accomplish the overall objective, the decentralized pursuit of subgoals calls for coordination. ∎

For business organizations, *coordination* can be defined as

> to establish and enforce rules which instruct organization members how to operate[8].

More generally, coordination refers to the rationales of *consistency*, claiming the compatibility in behavior of the contained subsystems, and *reliability*, meaning that the system's objective is pursued whenever each subsystem is functioning properly[9]:

Coordination is by definition concerned to ensure an effective interaction of system components. The degree of decentralization can range from an almost centralized system, over a distributed system with either central control or decentralized hierarchical control, up to a fully distributed system in which each subsystem is responsible for its own control[10]. A system coordination of the latter kind is e.g. supposed for socio-behavioral and biological systems. Since such systems are open in nature, it is verified that fully decentralized control can succeed.

Example 2.4 In the *model of Volterra* a simplified ecosystem consisting of predatory fish and prey fish is considered[11]. Volterra asked for the conditions necessary to reach a state of equilibrium in this *robber-victim* scenario. Accordingly, running into such a state is regarded as the systems objective in this context. In a first step the ecosystem is decomposed into two internal models describing the evolution of both populations separately. One assumes that the number of prey fish x increases at a constant rate $a > 0$ in the absence of predatory fish while the number of predatory fish y decreases at a constant rate $c > 0$ in the absence of prey fish. The growth parameters a and c represent the control in the decentralized subsystems. Obviously, if $x, y > 0$, both populations share a habitat, i.e. they will influence their evolution mutually. Thus one may assume that the prey fish population actually changes at a rate $a - by$ with $b > 0$ and that the predatory fish population changes at a rate $-c + dy$ with $d > 0$. In both models these internal transformations determine the feedback for the corresponding system control.

[8] cf. Arrow (1964).
[9] cf. Stadtler (1988).
[10] cf. Hynynen (1988).
[11] see e.g. Hofbauer and Sigmund (1984) for a comprehensive description.

Solving the resulting nonlinear equation shows that the ecosystem has two states of equilibrium, one trivial at $(x,y)=(0,0)$ and the other one at $(\overline{x},\overline{y})=(\frac{c}{d},\frac{a}{b})$. In the latter state the population sizes periodically oscillate to the mean values $\overline{x}$ and $\overline{y}$ respectively. By further analysis Volterra proved $(\overline{x},\overline{y})$ to be ***globally stable***, i.e. for arbitrary $x,y>0$ the dynamic system always runs into this state of equilibrium. Thus, under ideal conditions, decentralized coordination works well in order to keep the described ecosystem alive. ∎

In the above example the decomposition approach leads to an adequate understanding of the behavior of a natural system. By system analysis it is shown that actually no external coordination is needed to demonstrate the reliability of the fully decentralized system. Nevertheless, in the remainder of the thesis we will consider *self-coordination* as an extreme which is only claimed under aggregated assumptions made in analytical models.

Having discussed the difficulty of system decomposition quite generally, we now concentrate on the demands of artificial goal-seeking systems. Here, the most crucial aspect of problem decomposition is to design an adequate coordination for the resulting subsystems. In a computing environment this task has two dimensions. First, how is the exchange of information between adaptive systems realized, and second, how is the exchange of information controlled? We have answered the former question already in Sect. 1.3. The latter question asks for the strategy that governs coordination. In the following we show that diverse coordination strategies (independent of their degree of decentralization) are suitably represented within a hierarchical system framework. The information exchange is realized by system interventions where a subsystem is given the *right of intervention* into certain other subsystems. In this context, the affected subsystems are considered as hierarchically subordinate.

2.2 A FRAMEWORK FOR COORDINATION

In this section we introduce a framework capable to capture different strategies of coordination. We suppose a complex control system which has been decomposed into a number of decentralized subsystems. If only one of these subsystems contains a control unit, the overall coordination is certainly centralized there. In this case all other subsystems perform just transformation tasks under a central authority. If more than one control subsystem can be identified, the coordination is conspicuously decentralized. This raises the question of how the system coordination is conceptualized.

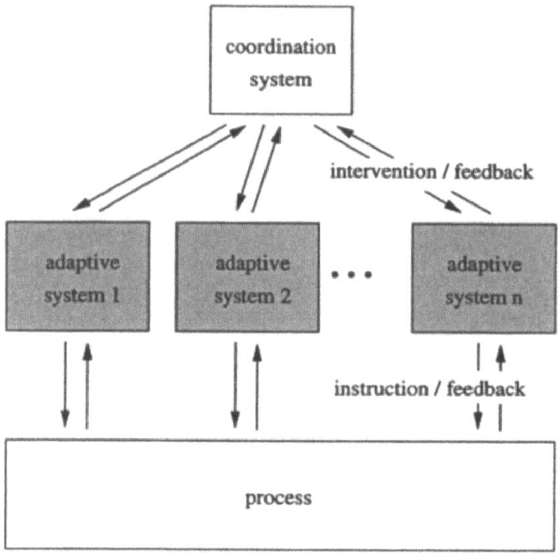

Figure 2.4 Structure of a two-level hierarchical system.

A frequently used way is to organize coordination according to the metaphor of a hierarchically structured business firms. There, decomposition identifies a distributed system with a decentralized hierarchical control. A comprehensive framework for coordination in multilevel hierarchical systems has been developed by Mesarović et al. (1970). A simplified variant of their approach is sketched in Fig. 2.4. Here the coordination task is related to a two-level hierarchical system. Considering this structure is sufficient because it can be used for integrating more complex coordination strategies too.

At the *top level* we face a system responsible for the coordination of n adaptive subsystems. Since each of these *base-level* systems contains a control unit the overall control is conceptually decentralized. Notice that all component relationships are vertically oriented. Accordingly, downward directed arcs either correspond to top-level interventions or to base-level instructions. In reverse, upward directed arcs represent base-level response or process feedback, respectively.

The structure of the coordination framework is dominated by a multilevel arrangement of control and controlled components. Since there exist only vertical component relationships, intra-level communication is ignored. Notice that the coordination system cannot access the process. Bottom-up relationships are used to response information gathered on deeper levels to upper levels. Consequently, intra-level communication can be realized by transmission through higher levels. This is an important feature for the coordination of decentralized subsystems. Here, the coordination merely manages an effective information exchange.

Due to the underlying metaphor, the coordination framework provides an appropriate interpretation of static organizations. We may look for instance at an organization with focus on its internal or its external behavior. From the inside view the organization is depicted as a multilevel system of interfering organization members. Hierarchically subordinate members are employed for problem solving via top-down relationships and report on their success via bottom-up relationships. From the external view the organization appears as a highly aggregated feedback-control system. Non members typically identify organizations with its purpose while any details of internal problem-solving remain unknown. External views therefore represent a suitable way to aggregate organizations which nest inside larger organizations.

2.3 ANTICIPATION

The intervention-feedback scheme provides an important element for coordinating a decentralized system in a conceptually hierarchical fashion. For goal-seeking systems, a further relationship between the various levels, known as *anticipation*, is often a helpful or necessary feature of an adequate coordination.

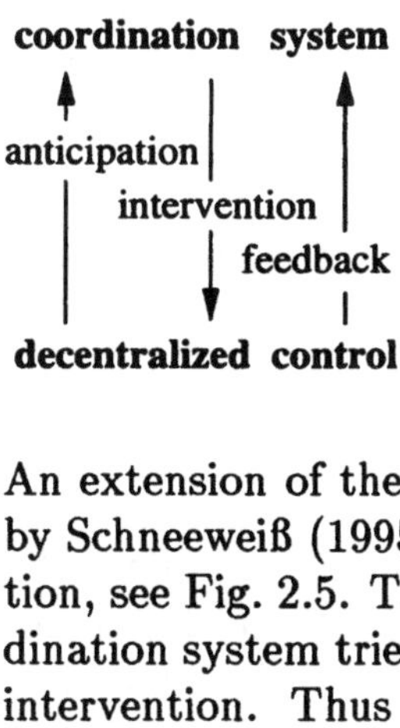

Figure 2.5 Incorporating anticipation into the coordination system.

An extension of the intervention-feedback scheme is therefore proposed by Schneeweiß (1995) which explicitly takes anticipation into consideration, see Fig. 2.5. Through anticipation of subordinate control the coordination system tries to include possible consequences into a hierarchical intervention. Thus anticipation is viewed as a bottom-up relationship which can precede a top-down intervention.

Anticipation is considered as an important mechanism involved in the related concepts of *hierarchical planning* and *hierarchical algorithms.* Roughly spoken, hierarchical planning is imposed to break down a system's complexity in order to be able to manage it. Accordingly, hierarchical algorithms effect the reduction of computational load in order to solve a planning problem. Together, hierarchical planning and hierarchical algorithms establish a methodological basis for anticipating goal-seeking systems.

3. DECISION MAKING SYSTEMS

In the preliminary investigations we have already used goal-seeking as a metaphor for decision making. Presupposing a formalized type of goal-seeking, both concepts actually base on the same rationale. Like goal-seeking, decision making can be regarded as a search process within a state space. In this section we present some examples for distributed decision making systems which address the coordination framework formulated above.

3.1 HIERARCHICAL AGGREGATIONS

In order to sketch frequent ways of organizing decision making processes we concentrate on a two-stage decision problem. Let us denote the decision made at the top-level as the outcome of Phase 1 and the decision made at the base-level as the outcome of Phase 2. Usually such kind of hierarchical decision making process is structured with respect to one of the following aggregate quantities[12].

State-based hierarchy. An aggregate decision is attained in Phase 1 which determines a global state for the final decision. One may think e.g. of a decision which leads to a number of mutually dependent subordinate decision problems. This set of problems is divided into a number of subsets. In Phase 2 the aggregate decision is disaggregated by solving the decision problems of each subset independently from the other subsets.

Objective-based hierarchy. If decision making deals with more than a single objective it is usually possible to derive a hierarchy for the objectives involved. Then, in Phase 1, a decision is made with respect to the most important objective. In subsequent phases this decision can be revised according to subordinate objectives.

Time-based hierarchy. Decisions which have to be made can effect the environment in long-terms. Hence it is often useful to anticipate consequences of potential decisions for a number of future time periods. In Phase 1 decisions with acceptable consequences are separated from those with unacceptable consequences. In Phase 2 a decision is selected from the acceptable ones which also appears favorable regarding short-term aspects.

Information-based hierarchy. The information required to make a certain decision is often not available or simply too complicated to take

[12] cf. Mesarović et al. (1970), Kistner (1992), Schneeweiß (1995).

it into account at a whole. A common remedy is seen in making a decision based on aggregated information available in Phase 1. Afterwards this decision is refined in Phase 2 with respect to more detailed information.

Hierarchical decision making often includes more than a single of the above mentioned quantities. For instance time-based approaches typically require the use of aggregated information at the top level. Since this information is often vague, detailed information is only available at the base level. Eventually the base-level control has shorter reaction times. Similarly, aggregate decisions in state-based approaches are frequently made with respect to preferences which are implicitly derived from a certain hierarchy of objectives.

We are now able to envision an abstract decision making process which is organized according to the framework sketched in Fig. 2.4. Every decision made at the top level represents a top-down intervention which, in turn, establishes the decision spaces for the base level. Unlike the top level, the base level deals with decisions that are implemented in the underlying object system. Thus, if the overall objective is attained, this stems from the decisions made at the base level. The most striking feature of such kind of process is that actually no system component has to pursue an overall objective. Since an overall objective is essentially outside of the system, it must be through coordination that distributed decision making can work effectively.

3.2 COORDINATION PRINCIPLES

In Sect. 1.3 we have discussed the structural prerequisites for the coordination of adaptive systems. Two types of hierarchical interventions, namely goal modification and image modification, have been distinguished. On this level of abstraction nothing is said about how coordination really influences the behavior of subordinate control. Recall that coordination has to take care for the consistency and the reliability of the entire system. Therefore coordination plays a crucial role in understanding multilevel decision making.

In the following we propose a classification of coordination principles which explicitly reflects the decision space of the coordination task. We outline three possible ways to coordinate a complex decision making process.

SUCCESSIVE COORDINATION

For each step of the coordination process, the decision space contains a single element, i.e. the coordination is completely determined. Typically, this way of coordinating a system consists in a transformation of feedback received from the base level into the next base-level instruction. During the transformation, aggregated quantities of time or information are processed in more detail. Thus successive coordination corresponds to a stepwise disaggregation of the overall decision process.

Example 2.5 We consider a capacitated variant of the well known model of Wagner and Whitin (1958). In this inventory-production problem it is assumed that an integer demand d_t of a single product is required in period t for a finite time horizon $t \in \{1, \dots T\}$. In order to satisfy the period demand d_t, units of the product can be produced in period t, they can be taken from a stock inventory if available, or alternatively, the demand is partially produced while the remaining demand is taken from the stock. The production capacity of each period is sufficient to produce at most Q units. The costs of production activity in period t are assumed to be a function of the quantity u_t, also called the *lot-size*. Since the demand must be satisfied in all periods, it can be favorable or even necessary to produce to stock in some periods. Keeping a quantity of s_t products on stock at the end of a period leads to inventory proportional holding costs. Thus the total costs incurring in period t are given as a function $c_t(u_t, s_t)$. Furthermore it is supposed that there is no entering inventory, and it is desired that there is no ending inventory. In this problem decision making refers to finding a sequence $[u_1, \dots u_T]$ which minimizes the cost function

$$f = \sum_{t=1}^{T} c_t(u_t, s_t). \tag{2.3}$$

The construction of the corresponding decision space is straightforward. For all $t \in \{1, \dots T\}$ the following conditions must hold[13].

$$\begin{aligned} & u_t \leq Q, \\ & s_t = s_{t-1} + u_t - d_t, \\ & s_0 = s_T = 0, \\ & u_t, s_t \geq 0, \text{ and integer.} \end{aligned} \tag{2.4}$$

Due to the integer constraints the decision problem is hard to solve in general. However, it can be decomposed into T subproblems which reflect the decision situation in each of the consecutive periods. From the balance equation we obtain $c_t(u_t, s_t) = c_t(u_t, s_{t-1} + u_t - d_t) = c_t(u_t, s_{t-1})$.

[13]The second equation is the *inventory balance equation*. Since d_t is given for all periods, each particular period decision u_t is equally represented by s_t.

The costs incurring in period t solely depend on the lot-size u_t and the inventories s_{t-1} which was hold in the previous period. This inventory level is known after the end of period t–1. At the beginning of period t we are interested in the remaining cumulated costs for $T-(t-1)$ more periods to go. These costs, denoted as $f_t(s_{t-1})$, result from the period costs c_t and the cumulated costs f_{t+1} incurring in the succeeding periods of t. Therefore it is a good deal to decide on u_t in every period by minimizing the resulting period costs c_t with respect to the subsequent cumulative costs

$$f_t(s_{t-1}) = \min_{u_t} \{c_t(u_t, s_{t-1}) + f_{t+1}(s_t)\}, \qquad t = 1, \ldots T. \tag{2.5}$$

From this recursion we directly obtain a successive coordination for solving the dynamic decision problem within a two-level system, see Fig. 2.6. Notice that the coordination is oriented in opposite direction to the enumeration of periods. The last period T is considered at first. Since there shall be no ending inventory $(s_T = 0)$, no cumulative costs incur after the last period $(f_{T+1}(0) = 0)$. The coordination system intervenes into the decision process in period T by prescribing the ending inventory $s_T = 0$. On this basis a vector of factual-optimal decisions concerning the possible inventories $s_{T-1} = 0, 1, \ldots d_T$ can be calculated from (2.5). Notice that an inventory $s_{T-1} > d_T$ cannot lead to the formerly prescribed condition. The corresponding vector f_T of cumulative costs for period T is responded to the coordination. Here f_T is transformed into an analogous instruction for the decision process in period $T-1$.

Continuing this way, the coordination finally reaches period 1. For this period a definite decision u_1 can be made with respect to the factual-optimal decisions of period 2 and their related cumulated costs f_2. The corresponding value of f_1 $(= f)$ is responded to the top level in order inform the coordination about the incurring inventory-production costs for the entire time horizon. The coordination task is completed then and the base-level systems consecutively make up their definite decisions u_t. Thereby they receive the inventory s_{t-1}, hold in the preceding period, from the underlying process. ∎

The system described in the above example actually generates optimal lot-sizes with respect to a finite time horizon of T periods. Roughly spoken this is due to a backward directed coordination of decision making. Factual-optimal decisions are generated in period t for $T-t$ more periods to go. Of course, this is a key strategy which is well known from Dynamic Programming. There it is called the ***principle of optimality***, see Bellman (1957). More generally one can argue that the struc-

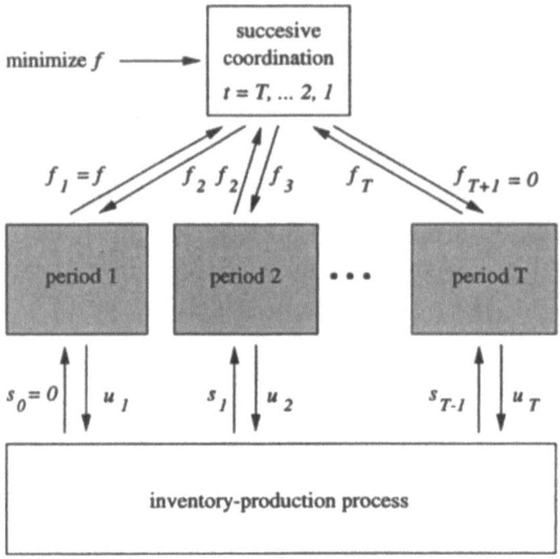

Figure 2.6 Successive coordination of an inventory-production system.

tural properties of Dynamic Programming have suitable interpretations through successively coordinated hierarchical systems[14].

GOAL-DIRECTED COORDINATION

If the decision space for coordination contains more than just one element, coordination has a direct influence on the overall system performance. In some cases an appropriate coordination of base-level activity is achieved by mutual adjustments with the top-level.

Example 2.6 Again, we consider an inventory-production problem. In extension of Example 2.5 this time N different products are subject to a given demand d_{it} for product $i \in \{1, \ldots N\}$ in period $t \in \{1, \ldots T\}$. In order to take different processing requirements of the products into consideration, product i is assigned a characteristic duration q_i which is needed for processing a single unit. Furthermore we assume that the time span available in each period is limited by a maximum of Q time units. Like before, the costs incurring for product i in period t are given by $c_{it}(u_{it}, s_{it})$, where u_{it} denotes the production quantity of product i and s_{it} denotes its ending inventory in period t. For lot-sizes $u_{it} > 0$ the production costs typically consist of fixed and of variable costs whereas the inventory holding costs for s_{it} units per period are usually considered as variable. For our purpose we neglect these details and formulate the objective as to find a sequence $[u_{11}, u_{21}, \ldots u_{NT}]$ which minimizes the

[14]For an outline and further examples see Schneeweiß (1994).

cost function

$$f = \sum_{t=1}^{T} \sum_{i=1}^{N} c_t(u_{it}, s_{it}). \tag{2.6}$$

According to (2.4) the corresponding decision space is defined for all products i and all periods t by the following conditions

$$\begin{aligned} &\textstyle\sum_{i=1}^{N} u_{it} q_i \le Q, \\ &s_{it} = s_{i,t-1} + u_{it} - d_{it}, \\ &s_{i0} = s_{iT} = 0, \\ &u_{it}, s_{it} \ge 0, \text{ and integer.} \end{aligned} \tag{2.7}$$

This optimization problem is called *Capacitated Lot-Sizing Problem*[15] (CLSP). By taking a closer look at the CLSP it strikes that production is modeled like a single-stage process. It is presupposed that the capacity Q can be fully exploited within each of the periods by the first constraint in (2.7). Of course, this is an oversimplification which corresponds to the facts in case of single-stage production only. If production is a multistage process it turns out that the duration $u_{it}q_i$, required to produce a single lot, represents only an aggregated time span. Sequence-dependent intermediate delays which occur whenever a lot switches between two stages of production are not contained in $u_{it}q_i$. Since the exact time consumption is unknown on the level of lot-sizing, the aggregated capacity constraints are used. The computation of more exact values (i.e. a detailed anticipation of a period) requires to take a multi-stage model of the production system (e.g. a job shop) into consideration. This level is referred to as the level of production scheduling. Thus it turned out that inventory-production decisions are actually related to a combined problem of lot-sizing and scheduling. Therefore production planning in total can be treated as a hierarchical process.

In the standard top-down approach lot-sizes u_{it} are computed by solving the CLSP or an alike lot-sizing problem at first. Then, for each period involved, one searches for a production schedule compatible with this plan. In general production scheduling can pursue different objectives. Since time is the scarce resource in this example we have to concentrate on finding a detailed schedule with respect to the lot-sizes u_{it} in period t that can be realized within Q time units. This objective

[15]The CLSP is known to be an *NP*-hard optimization problems in almost every case. Therefore heuristics must be used in order to solve larger instances efficiently, e.g. the well known heuristics of Dixon and Silver (1987) or Günther (1987). However, in contrast to Example 2.5 its not our intention here to show how any of such algorithms work. We are rather interested to see the complex relationships resulting if the decision problem at hand is coupled with another one which is viewed as hierarchically subordinate.

addresses the minimization of *makespan*, i.e. the time span needed to make the products planned so far. Of course, if we find a schedule with a makespan lower than or equal to Q we say that a schedule is compatible. Recall that in order to gain consistency for the entire system, the mutual compatibility of subsystems is required. Often, however, it will be difficult or even impossible to find compatible schedules for all periods. Consequently, the demand cannot be satisfied in at least one period t. Since backlogging of demands is not allowed, units of products which cannot be completed within that period must be shifted into periods $t' < t$. In other words, the original plan derived from lot-sizing is revised in a way which directly increases the costs of inventory holding.

Thus far we have described a two-level decision making system for integrating lot-sizing and scheduling. Although the interpretation of components and relationships is clearly different, its structure is similar to the one sketched in Fig. 2.6. At the top level we face a coordination system prescribing lot-sizes of products to the periods with respect to low inventory holding costs. At the base level detailed schedules are generated for each of the periods. Whenever a compatible schedule cannot be found, the capacity bottleneck is reported to the top level. Eventually the coordination resolves the bottleneck by shifting critical lots into earlier periods. Often trial-and-error revisions are used for this purpose. More sophisticated procedures alternate between the two levels until consistency is achieved[16]. ■

In this example coordination essentially manages the elimination of capacity bottlenecks. Decisions concerning the way how to eliminate a bottleneck can always be accessed in terms of a costs function. Consequently, coordination cannot concentrate on system consistency alone. From a reliable coordination of the inventory-production process we expect to take care for moderate inventories as well. For other problems an unbiased coordination appears more suitable because there are no coordination-dependent costs or they are simply ignored. We consider this case at last.

INDEPENDENT COORDINATION

The coordination task is problem independent, i.e. the effect of top-level decisions cannot directly be measured in terms of an objective function. Consequently, reliability concerning the accomplishment of an objective must be subject to the base-level control. The top level basically serves as a platform for exchanging base-level messages and

[16] cf. Dauzère-Péres and Lassere (1994).

solving base-level conflicts. In order to attain impartial decision making, the coordination can e.g. follow a prescribed scheme of conducting negotiations.

Example 2.7 We consider a logistic model, the so called single-depot *Capacitated Vehicle Routing Problem*[17] (CVRP). In this problem a number of N customers has to be served by a fleet of K identical vehicles of limited transport capacity Q. The distance between two customers $i, j \in \{1, \ldots N\}$ is known and denoted as d_{ij}. Accordingly, the distance between customers and the central depot is denoted as d_{0i}. For simplicity we suppose symmetric distances between all locations. Serving customer i exhausts $q_i \leq Q$ units of transport capacity of a vehicle. In order to ensure the existence of feasible solutions we assume that the total capacity available is clearly larger than the capacity needed. Typically, not all vehicles of the fleet are used to serve the customers. Initially, the entire fleet is located at the depot. The objective is to find a set of K routes (one for each vehicle) which minimizes the total distance

$$f = \sum_{k=1}^{K} \sum_{i=0}^{N} \sum_{j=0}^{N} x_{kij} d_{ij}. \tag{2.8}$$

Here, x_{kij} is a binary decision variable denoting whether vehicle k visits customer j immediately after customer i (x_{kij} =1) or not (x_{kij} =0). Each route begins at the depot, then visits a (not necessarily non-empty) subset of customers without violating the capacity constraint and finally returns to the depot. For our purpose it is not necessary to formalize the CVRP in greater detail.

Traditionally, there are two classes of methods solving the CVRP in a hierarchical fashion. One class consists of those methods that first construct an efficient route through all the customer's locations and then partition this route into segments respecting the capacity constraint. That way, each segment is assigned to one vehicle which visits the customers consecutively according to their order given on the route. Methods found in the other class proceed in reverse direction. Here customers are first clustered into feasible subsets which are served by the same vehicle. Since no ordering has been specified before, efficient routes have to be searched for the vehicles afterwards. Both concepts provide a problem decomposition leading to significant increases of the overall complexity. None of both classes, however, contains methods that guarantee optimal

[17] Like the CLSP this problem also belongs to the class of *NP*-hard problems, see Bramel and Simchi-Levi (1997) for an overview on vehicle routing models.

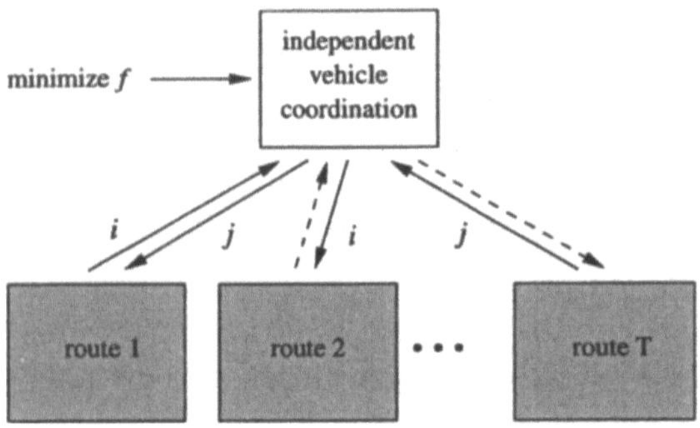

Figure 2.7 Independent coordination of fleet management through customer exchanges.

overall decisions because it is by no means clear that the decomposed problem space contains optimal solutions anymore. Thus, even if exact algorithms for clustering and routing are applied, it is rather unlikely that the best overall decision will be made.

A remedy is promised from the exchange of customers between vehicles. Fig. 2.7 shows this idea in terms of the two-level coordination framework[18]. Exemplary, customers i and j are exchanged between vehicles 1, 2, and T. Each customer location can be valuated in terms of route-specific visiting costs. E.g. for route 1 we calculate the costs incurring if customer i is visited in between of customers p and q by $\Delta_i = d_{pi} + d_{iq} - d_{pq}$. Since the triangle inequality holds for symmetric distances, Δ_i expresses a non-negative saving gained for route 1 if customer i is served by some other vehicle. Customers yielding above-average savings are promising candidates to be dropped from a route. Accordingly, customers yielding below-average savings and which are currently assigned to other vehicles represent interesting candidates to be included in a route.

The exchange of customers between vehicles can be managed by introduction of a virtual market[19]. In this context the components of Fig. 2.7 are interpreted as autonomously acting entities which are referred to as *agents.* At the base level we place *route agents* which determine a route for a vehicle visiting its customers. Less profitable costumers are offered by the route agents at the exchange market. Every offer is valuated by the other route agents in the context of their own routes. Typically,

---

[18]For reasons of clarity the associated process has been left out this time. We can think e.g. of an object system providing the bottom-level with exogenous data like the distances between locations.

[19]This concept is known by the term *Simulated Trading*, cf. Bachem et al. (1992). The virtual exchange market is generally organized as a *contract-net* of cooperating and/or competing actors. The communication of actors is monitored by a so called contract-net protocol which is initiated by an additional actor called the contract manager, cf. Smith (1980).

this requires to reorder the route planned for a vehicle. At the top level we have a further agent playing the *contract manager*. Its task is to coordinate a consistent exchange of customers between vehicles, i.e. all customers must be served and vehicles must not exceed the allowed capacity. Since every customer exchange is accessed in terms of the objective function (2.8) by the route agents, the minimization of total distance must not be pursued by the contract-manager. If several route agents compete for the same customer, the contract manager may decide at random. In more sophisticated variants, the route agents are requested for bids representing their expected savings. Then the conflict can be solved according to economic rationales, i.e. by means of the bidding scheme implemented in the contract-net architecture[20]. ∎

In the above example the base-level systems can act autonomously to a large extent. The relationships existing with the top-level do not really obey to the intervention-feedback scheme. The coordination can request a base-level system but it may be the other way round as well.

Many problems permit various ways of coordinating a decision process. For the CVRP recent advances have been made with a centralized coordination of sophisticated customer-exchange techniques[21] which is heuristically controlled by the principle of *local search*. For other problems a totally decentralized control may succeed where the base-level components request each other in a one-to-one fashion using the top-level just to realize intra-level communication. The general aim of such approaches is to describe interrelated activities which can be performed by agents in a real-world setting[22]. The formal coordination principles established in such approaches are usually adopted from well-known market-mechanisms, e.g. a particular kind of auction.

3.3 DISCUSSION

Since the necessity of coordination can obey to different rationales we have formulated three principles which fit the needs for a wide application range.

The principles presented to solve the CLSP and the CVRP show noticeable similarities. Both problems deal with a limited resource for which an efficient allocation plan is searched. In both cases we face several interdependent sequencing problems at the base level, one time related to scheduling and the other time related to routing. Accord-

[20] see Zelewski (1993) for a comprehensive discussion of bidding schemes in a contract-net based process coordination.

[21] cf. Kindervater and Savelsbergh (1997).

[22] cf. Unland et al. (1996).

ing to the definition of aggregate quantities given in Sect. 3.1, decision making operates upon a state-based hierarchy in both cases. In spite of this there is an extremely important difference between both planning models. For the CLSP every decision made at the top level leads to additional costs which cannot be taken into account at the base level. Goal orientation is therefore a crucial attribute of coordination. As we have observed for the CVRP, an independent coordination may succeed which does not take the overall objective into account.

The principles of goal-directed and independent coordination appear related if compared with a successive coordination. The most evident difference results from the prefixed number of coordinating steps carried out. Often this coordination principle rather corresponds to a hierarchical algorithm than to a hierarchical concept of planning.

Our coordination framework differs slightly from other schemes proposed in literature. Approaches in multilevel planning are often distinguished with respect to the relationships existing between the hierarchical levels of decision making. Significant differences of coordination can become visible if we focus on the incorporation of anticipation, intervention and feedback functions. For instance Schneeweiß (1994) refers to multilevel decision making without anticipation and feedback as *successive planning*. Similarly, Kistner (1992) defines *hierarchical planning* as successive planning using feedback mechanisms for coordination.

Successive planning. In order to derive an intervention, the top level does not anticipate the base level explicitly. There is either no anticipation at all or anticipation is restricted to key features of the base level which are not further disaggregated. Base-level feedback is ignored.

Hierarchical planning. The top level anticipates the base level by using aggregated data which is subsequently processed in higher detail. Typically base-level feedback leads to readjustments of the top-level control system, but it cannot trigger a revision of the top-level interventions.

Integrated planning. Insufficient base-level anticipation is diminished by alternately solving base-level and top-level decision problems. Base-level feedback provides a more detailed anticipation at the top level which may lead to revisions of the previous intervention at the same time.

Notice that our three coordination principles do not exactly match the above scheme. We may feel inclined to equate successive planning and successive coordination although successive coordination may be addressed to hierarchical planning as well. Example 2.5 shows how anticipation and feedback can be incorporated into a stepwise coordination of a hierarchical algorithm. A famous example for hierarchical planning is known by the aggregate production planning model of Hax and Meal (1975). Integrated planning, on the other hand, is closely related with goal-directed and with independent coordination, as we have verified by examples 2.6 and 2.7.

To summary, various work has been addressed to multilevel decision making systems. The connecting elements of many approaches become evident if they are projected on their internal feedback-control structures. Despite that, multilevel decision making still is a conceptual approach more than a strict recipe. However, from this detached view of general systems only problem solving techniques and their coordination need to be specified.

4. SUMMARY

In this chapter we have borrowed the terminology of general system theory in order to raise issues belonging to adaptation, complexity and coordination. The advantage promised is the following. First, we hoped to get insight into the structure of planning problems addressing the management of logistics systems. Second, we hoped to get insight into the structure of adaptive problem solving. Finally, while correspondence is reached for both fields on the description level, we hope to attain more easy compatibility on the processing level too.

In order to put things in concrete terms some examples of two-level decision making processes have been considered. One alternative to solve such problems at the base level is to use adaptive search algorithms for optimization. After having outlined the general structure of adaptive systems in this chapter, we are able now to integrate such algorithms in a complex search. For the coupling of adaptive algorithms on several levels of search we can take advantage from the principles of system interventions. What is still needed is an internal view of adaptive search methods and their problem-solving capacities.

Chapter 3

GENETIC ALGORITHMS

Having focused on the history of evolutionary computation in Chap. 1, we discuss the paradigm in greater detail within this chapter. As mentioned before, the class of EAs contains several substantially different approaches. Probably most famous are Genetic Algorithms (GAs). In this chapter we concentrate on this mainstream.

At current time GAs are widely used in optimization and in other applications for at least one of two main reasons. One is their large success in many fields and the other is their comparable ease of implementation. Since the GA paradigm was founded a lot of research has been devoted to improve the basic concept. In the first section of the chapter we shortly revisit the basic concept. Then we turn to the most frequently used modifications and extensions of GAs in the second chapter. Much of this work is based on arguments of plausibility and on computational experience. In contrast, theoretical results on the justification of GAs are relatively scarce although much effort has been spent in this field. We present two such attempts in the last section.

1. BASIC PRINCIPLES

Genetic Algorithms are supposed to work on a population of binary strings which are also called chromosomes in the GA lingo. Each string encodes a certain solution of the problem space to be searched. Since solutions are encoded in binary form, a GA can employ context-free search mechanisms on a syntactical basis.

In order to guide the syntactical search mechanism in the boolean space, there is given a function f which measures the *fitness* of binary strings. The fitness function is derived from the objective function of the problem under consideration. Usually the fitness function is maximized and assumed to be non-negative. If, however, the objective function has to be minimized or the function happens to take negative values, then the fitness function must be obtained from a transformation of the actual objective function.

1.1 POPULATION MANAGEMENT

The binary strings correspond to the individuals of a natural system. A preset number μ of such individuals is supposed to form a population which is maintained in parallel by a GA. Initially, the population is generated at random. We denote this state of the population as P_0 and, accordingly, the state achieved after t time steps as P_t.

Any GA basically follows the iteration described in the reproductive plan of Holland (1975, 1992). The transition from P_t to P_{t+1} is executed in three steps. First, each individual of the population is decoded into a candidate solution and evaluated in terms of the fitness function. Second, μ individuals are selected from P_t with respect to their fitness and copied into an intermediate population P^*. Finally, offspring are recombined from the individuals in P^* by applying operators modeling processes of natural reproduction. These offspring replace the previous population.

```
algorithm GA is
    initialize P(t:=0) at random
    while not terminate do
        evaluate individuals of P(t)
        select individuals from P(t) to P*
        recombine individuals from P* to P(t:=t+1)
    end while
end algorithm
```

Figure 3.1. Outline of Genetic Algorithms.

The transition from P_t to P_{t+1} corresponds to a progress of one *generation* in the simulated evolution. Typically, many generations are carried out until a *termination criterion* is met. Most widely used is a preset number of generations to be iterated.

According to Angeline (1995), the transition performed within one generation is expressed by

$$P_{t+1} = \varphi(P_t, f(P_t)). \tag{3.1}$$

Here, the particulars of the transition function φ determine the mechanisms for selecting and producing individuals. Selection as well as reproduction are responsible for generating improvements in terms of the objective pursued. Therefore these topics are subject of the following consideration.

1.2 SELECTION MECHANISM

In biology the fitness of an individual is simply defined by the number of offspring of the individual. On average, fitter individuals do have more offspring. In simulating evolution we proceed just the other way round: instead of determining the fitness of an individual by its number of offspring, we calculate its fitness by means of the fitness function at first, and produce a corresponding number of offspring afterwards.

The fundamental principle of evolution, known as *natural selection*, is implemented in the GA by selecting individuals from P_t to P^* in proportion to their fitness. The average fitness of P_t is calculated by $\overline{f}_t = \frac{1}{\mu}\sum_{x \in P_t} f(x)$. Then each member of P_t is assigned the normalized fitness

$$p_x = \frac{f(x)}{\overline{f}_t}, \tag{3.2}$$

expressing its probability to be selected for reproduction at one draw. The mechanism, called *proportionate selection*, subsequently carries out μ such "draws with replacement" out of P_t, until the intermediate population P^* is filled. This technique presents a conspicuously close approximation of natural selection because an individual of above average fitness is given a reasonable higher chance to have offspring.

1.3 REPRODUCTION MECHANISM

In evolutionary genetics, reproduction stands for a copy process of genetic information which is passed on from the parents to the offspring. The evolution of a population is defined as the change of the gene frequencies given in the *gene pool* of the population. From this point of view the individuals merely function as carriers of encoded information.

Selection already changes the gene frequencies of the gene pool, but does not produce "new individuals" with respect to their genetic information. This task is performed by a recombination process which is disturbed by coincidental effects on the gene recombination.

The central principles of genetic transmission are implemented in a GA in terms of operators which recombine new individuals from the information stored in the selected parental strings. A simulation model of natural reproduction is defined by a set of operators and a corresponding set of probabilistic control parameters. First, the intermediate population (containing μ parental strings) is split into $\mu/2$ arbitrary pairs. Next, a so called *crossover operator* is applied with a preset probability, often ranging in between of 0.6 and 1.0. The standard crossover operator, referred to as *one-point crossover*, exchanges the bit values to the right of a randomly chosen *crossover-point* between a parental pair of strings.

parent 1	0	1	1	0	0	0	1
parent 2	1	1	0	1	1	1	1
offspring 1	0	1	1	0	1	1	1
offspring 2	1	1	0	1	0	0	1

Figure 3.2 The one-point crossover operation.

The idea of this procedure is to generate substantially new individuals which may show an improved fitness. If the crossover operator is applied, the resulting strings are copied into population P_{t+1}. Otherwise the strings of a parental pair are copied directly from P^* to P_{t+1}.

In order to model slight copy-errors of reproduction the *mutation operator* may randomly invert some bit values while the strings are copied. A mutation can (re-)introduce a bit value to the gene pool of P_{t+1} which did not exist in P_t. Too many mutations, however, result in a random search and therefore a small mutation probability of 0.001 or less is used for each bit copied. Further genetic operators such as *inversion* and *translocation* have been investigated by Holland (1975, 1992) as well, but they are hardly used in GA practice.

Thus far we have outlined the basic principles employed for searching a boolean space. To summarize briefly, the GA performs a population-based search by maintaining a multitude of candidate solutions in parallel. In order to improve the average fitness of the population, the most promising solutions are given the highest selection rates. The intention of reproduction is to generate increasingly better individuals on the basis of a continuously modified gene pool.

2. MODIFICATIONS AND EXTENSIONS

Research on GAs has received much attention in the last decade. In order to enhance GA performance, several modifications and extensions of Holland's reproductive plan have been developed. Consequently GAs have a rich literature including several textbooks as well. Since the field is increasingly active the related literature rapidly accumulates. A comprehensive coverage seems hardly possible within this scope, thus we restrict ourselves to a discussion of the most remarkable approaches.

Among the textbooks the classical work of Goldberg (1989) is certainly the most widespread one. A collection of early papers dealing with a lot of practical design aspects of GAs is collected by Davis (1991). Other introductions to the field are given by Liepins and Hilliard (1989) and Michalewicz (1996). A frequently met shortcoming of the early literature is the euphoric treatment of the biological analogy of GAs. Since limitations of this metaphor have not been recognized at that time, the basic tenor often appears immaturely today. A modern book which provides a more critical appraisal of the paradigm is due to Mitchell (1996).

2.1 ALTERNATIVE POPULATION MANAGEMENT

A bulk of research focuses on alternating the population management. Contrasting the base iteration shown in Fig. 3.1, various strategies have been developed, avoiding to replace all individuals in every generation. A simple case is the *elitist strategy* which always transfers the best individual into the next generation. The approach aims more on the improvement of one promising individual than on increasing the fitness of the population in average. Another idea is to carry out a combined selection-recombination step in each generation such that offspring immediately replace their parents. In this way the population remains in a *steady-state* which provides the selection of a newly generated individual already in the next step, see Whitley (1989).

Other approaches introduce a population structure by distribution model for the individuals. Mühlenbein and Gorges-Schleuter (1988) place individuals in overlapping areas, called *demes*, in order to introduce spatial distances in the population. Similarly, Tanese (1989) divides a population into a number of distinct subpopulations, called *islands*. In both approaches selection and recombination are carried out locally such that the GA receives a decentralized control. The information flow is realized by diffusion through overlapping demes or by individuals migrating between islands. Distributed populations effect a spatial or a temporal isolation of the individuals. Isolation of individuals is expected to be helpful because different characteristics of individuals may evolve at the same time without competing immediately against each other in the selection step. Moreover, distributed populations are straightforward if a parallel implementation of the GA is intended, see e.g. Mühlenbein and Gorges-Schleuter (1988); Kopfer et al. (1996), and Dorigo and Maniezzo (1993) for an overview on this topic.

2.2 SELECTION TECHNIQUES

Research also concentrates on selection schemes guiding the search more properly than the above presented selection mechanism can. A weakness of fitness proportionate selection is observed if the best fitness value in a population falls towards the average fitness, compare (3.2). Let us consider a population where the best fitness is 101 while the average fitness is 100. In many applications this lead in fitness will indicate a substantial progress of search. Even so, under proportionate selection the fittest individual is given a chance for reproduction only 1% higher than on average. Several ways of remedy are proposed including *fitness scaling*, *tournament selection* and *rank-based selection*. A detailed description of these techniques is given by Beasley et al. (1993). For a comparison of various selection methods see also Goldberg and Deb (1991).

2.3 ENCODING AND GENETIC OPERATORS

The way of encoding candidate solutions is one of the most crucial factors for the success of a GA. Unfortunately there is no general guideline how to represent a particular problem appropriately. Since theory concentrates on binary encoding, many researchers follow this lead. An encoding of integers $x < 2^n$ is derived from the binary representation $x = \sum_{k=0}^{n-1} a_k 2^k$, with $a_k \in \{0,1\}$. On this basis arbitrary intervals of real-valued numbers can be represented as well with a prescribed precision.

Although straightforward, this binary encoding of a continuous problem space is often unfavorable, because solutions which are adjacent in the problem space possibly do not share a single bit in their binary representation. This applies e.g. for 0111 and 1000, where the strings encode the adjacent integers "7" and "8" respectively. As shown by Caruana and Schaffer (1988), the use of *Gray-code* is advantageous because it always maps adjacent points into strings which differ in only one bit position. Consequently, a single mutation can change "7" into "8". Thus, using Gray-code, the mutation operator can work on a reliable basis by effecting slight modifications in the problem space, as intended.

The suitability of an encoding scheme can be discussed from the perspective of the crossover operator too. Let us assume the GA has generated a new string in which the bit 1 is placed at two certain positions. Possibly, this string is of high fitness just because the two bits appear together. However, if both bit positions are far away from each other it is rather unlikely that one-point crossover (see Fig. 3.2) will not separate the bits again. Consider e.g. the string 1010. Only in case that the crossover point is chosen after the third bit position, the two bits are not separated by one-point crossover.

In order to reduce this *positional bias* of crossover, several slightly different operators have been suggested for searching the boolean space. Basically these operators increase the number of crossover-points involved in one operation. Widely spread in practice is *two-point crossover.* In *uniform-crossover*, a crossover point can happen at each bit position with a prescribed probability, see Syswerda (1989). Therefore uniform crossover has no positional bias: any bits located at different positions in one parent can potentially be recombined in the offspring. However, to decide whether or not positional bias is favorable for crossover requires to take the structural properties of the encoded problem space into a detailed consideration.

Regarding combinatorial order problems it turned out that a bit-oriented encoding of solutions is unfavorable because crossover leads to infeasible solutions in most cases. This inconsistency is avoided by using permutations for encoding candidate solutions. Consider a permutation like *abcd* which represents a possible order of the elements of an object set $\{a, b, c, d\}$. Using such kind of string representation, the GA actually searches a permutation space, resulting in feasible solutions in all cases. As already mentioned in Chap. 1(2.2), the approach is referred to as an order-based GA. Several mutation and crossover operators, preserving the permutation structure, have been developed for order-based GAs. For a survey see Fox and McMahon (1991).

2.4 CONSTRAINT HANDLING

Another problem closely related to the representation issue is the handling of constraints. If an individual cannot be transferred into a solution which satisfies all constraints, the encoded candidate solution obviously represents an infeasible point in the search space. A simple remedy is suggested by using *penalty functions.* A penalty function deforms the fitness function used by the GA such that selection can navigate search into feasible regions of the search space. Of course, the approach cannot succeed if the set of infeasible solutions is dense within the search space.

Typically one starts with relaxed penalties and tights them as the GA progresses. In the approach of Richardson et al. (1989), each constraints of a problem is assigned a certain weight. The weights play the role of penalties if a candidate solution violates them. The fitness $f(x)$ of an infeasible solution x is decreased by the corresponding weight. In order to increase penalties in later generations we can use a function like $P(t) = \frac{t}{T} \overline{f}_t \sum v_i$. Here T denotes the preset number of generations until the algorithm terminates and v_i denotes the violation weight of the i-th constraint.

2.5 DECODING

Often the constraints of a problem effect that each feasible solution is adjacent to many infeasible points in the search space. This difficulty applies e.g. for many combinatorial optimization problems. It was observed first by Davis (1985) when approaching order-based GAs for a scheduling problem. In order to represent schedules by permutations, Davis introduced the principle of *encoding and decoding*. Here, the information encoded in the permutation does not determine a particular candidate solution directly. An individual is rather considered as a recipe how to build a feasible schedule for a specific problem instance at hand. These particulars are hidden inside of a procedure which *decodes* the permutations.

A decoding procedure either interprets a string as an instruction list for building a feasible solution or it derives a solution syntactically and checks its feasibility afterwards. If infeasibility is detected, the procedure employs a repair method which transfers the infeasible point into a related feasible one. In both cases the decoding procedure access the detailed data of a problem instance. In the previous investigation this right was reserved for the fitness function. Although often not explicitly mentioned, the possibility of incorporating encoding-decoding procedures in GAs has become one of the most important techniques of evolutionary computation.

2.6 HYBRIDIZATION

Problem-specific knowledge can be incorporated in GAs in diverse ways. For instance we can seed the initial population with solutions produced by greedy heuristics or we can apply heuristic recombination operators. A comprehensive survey is given by Grefenstette (1987).

Like any other search paradigm, GAs have to balance between the exploration of new solutions and the exploitation of promising solutions already found. Due to the incorporation of crossover, GAs have attractive capabilities for exploration, but they are often rather weak in exploitation. In order to enhance effectiveness it is therefore straightforward to combine GAs with problem-specific gradient methods. In this way a local optimal solution can be reached from a starting solution previously generated by the GA.

This idea, referred to as *genetic local-search*, was first approached by Mühlenbein et al. (1988) and Ulder et al. (1990) in the domain of combinatorial optimization. Here, an iterative improvement algorithm is incorporated in the decoding procedure used for evaluating the individuals. Actually GAs are often applied just because no other efficient methods are known. However, if genetic local-search is possible it is interesting to aware that the role of the GA has changed to some extent. Now the GA is responsible for navigating another heuristic within the search space which takes over exploitation.

2.7 PARAMETER TUNING

Having made the design decisions for the GA we still have to choose a suitable parameter setting. Research has spent much effort on sizing populations efficiently, prescribing the number of generations for termination, and choosing rates for crossover and mutation optimally. A comprehensive overview is given by Mitchell (1996). Since all these parameters are mutually dependent, they cannot be optimized consecutively. Larger populations lead to a more thorough search at the expense of longer computation time needed. As shown by Nakano et al. (1994), there is a trade-off for using larger populations under constant computational cost. Regarding efficiency, it is therefore often favorable to use smaller populations while iterating more generations or running the algorithm several times. Similarly, the rates of mutation and crossover should interact for a careful balance between exploration and exploitation.

Currently there are no conclusive results on what works best. Many GA practicians use frequently reported parameter settings while others try to find the best setting by experiment. A recent discussion refers to the *no-free-lunch* theorem for search, saying, whenever an optimal parameter setting is found for a problem instance, there exists another instance of that problem class for which this setting is totally inadequate, compare Wolpert and Macready (1997). Under this presumption it would be advantageous if the parameters automatically adapt during the evolutionary computation. We deal with this topic later on.

3. THEORETICAL CONSIDERATIONS

An important issue addresses the question why GAs work at all. Of course, this question has been raised from the early beginning of evolutionary computation. Nevertheless, so far no answer has been given which is accepted by the whole research community.

To realize the difficulty let us recall that an EA is nothing more and nothing less than a very simple model of natural evolution. In reverse we can state that natural evolution is certainly not an optimization method. Natural evolution just evolves. For this reason it can be argued that the connection between natural and artificial evolution is not close enough to believe that any argument valid in one world is valid in the other one too. On the other hand it is also argued that design decisions should follow principles if these are found in nature as well. Accordingly, the success of a particular algorithm is explained by the care spent on modeling nature-like mechanisms. To simplify matters in this way certainly prevents a deeper understanding of the way how EAs really work. Therefore statistical models of the computation have received much attention.

3.1 STATISTICAL MODELS

The first of such models has been proposed already by Holland (1975, 1992) under the name *Schema Theorem.* Interpretations and implications of this analytical approach are still discussed intensively. For a comprehensive survey of consecutive developments in this field we refer to Goldberg (1989); Grefenstette (1993) and Mitchell (1996).

The theorem is basically described as follows: a GA works by discovering, valuating and recombining the *building blocks* of promising candidate solutions. The underlying idea is that solutions produced by a GA are composed from building blocks. The better the building blocks are, the better a solution can get. Since the fitness contribution of building blocks is implicitly estimated by the selection mechanism, above-average building blocks are sampled at higher rates, leading to increasingly better solutions. Formally, a building block corresponds to a hyper-plane of the boolean search space (referred to as a *schema* by Holland). Under proportionate selection and one-point crossover conditions, the processing of schemata can be analyzed statistically. Unfortunately, the results gained cannot be verified empirically in all cases[1].

[1] The relevance of the schema theorem has been questioned by Mühlenbein (1991) and Grefenstette (1993) with regard to so called GA-*deceptive* problems. As a consequence, a class of fitness functions, called *Royal Road Functions*, has been investigated which is expected to be most suitable for GAs. But the results achieved are disappointing since it turned out that gradient methods still can outperform a GA in this domain, see Mitchell et al. (1994).

Another much more comprehensive model of GA behavior has been suggested by Goldberg and Segrest (1987). Here it is argued that a GA performs a dynamic process. According to (3.1), this process iteratively performs a state transition. Since the fitness function does not change in time, the exact state at generation t is sufficiently represented by the population P_t. Hence the subsequent state P_{t+1} solely depends on P_t. Notice that the transition function is stochastic because of the various incidental events in selection and reproduction. Consequently, describing the GA behavior in detail requires to use a *Markov chain* where each state of the chain represents a corresponding population. This idea was taken up by other research leading to theoretical results concerning the capability of GAs to progress, see Vose and Liepins (1991). The practical use of Markov-chain analysis is seen in predicting the GA behavior. In principle, the approach allows us to derive appropriate GA parameters. Unfortunately, the application of Markov-chain analysis is limited to relatively small search spaces. Larger problems require larger string representations which rapidly increase the number of possible GA states. This makes the analysis intractably hard.

The statistical models described so far suffer from a lack of evidence when applied to problems with a deceptive structure or a non-trivial size. Furthermore the models are completely inadequate to predict the GA behavior whenever modifications or extensions of the basic principles are used. For instance the schema theorem cannot directly be transferred to order-based representations. Similarly, sophisticated ways of the population management are difficult to handle in a state transition process. For such reasons many GA practicians tend to stress the metaphor of natural evolution.

It is not our intention to justify or contradict the theoretical or the pragmatic view. Still, it cannot be doubted that EAs are based on the adoption of a few basic mechanisms which are also used for modeling natural systems in theoretical biology. Thus it may help to look how evolutionary processes are treated there.

3.2 POPULATION GENETICS

The dynamic behavior of GAs can be analyzed by applying quantitative methods developed in population genetics. This interesting idea is outlined in great detail by Mühlenbein (1997). He investigates the progress of a population of binary strings by means of the dynamic effected under proportionate selection. For this purpose the *selection differential* is considered, which calculates the difference between the mean fitness $\overline{f^*}$ of the selected strings in the intermediate population P^* and the average fitness. In generation t we observe the fitness differential

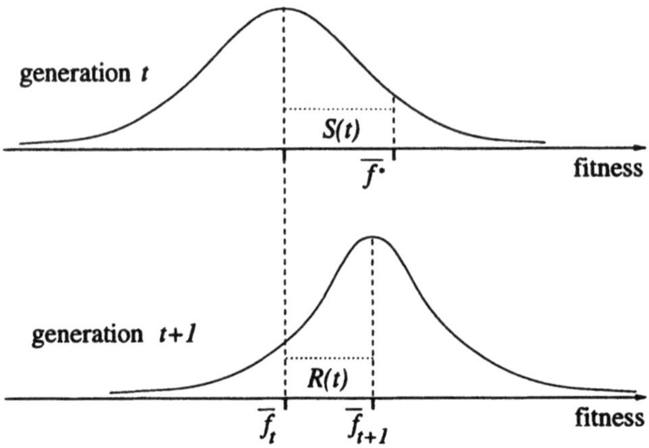

Figure 3.3. Transition of the fitness distribution from P_t to P_{t+1}.

$S(t) = \overline{f^*} - \overline{f}_t$. This quantity can be used to predict the improvement of the average fitness in generation $t+1$. Formally it is expressed by $R(t) = \overline{f}_{t+1} - \overline{f}_t$, where R is referred to as *response to selection.* The connection between $S(t)$ and $R(t)$ is illustrated in Fig. 3.3. As common in population genetics, it is assumed that the fitness of individuals is normally distributed within the populations.

Typically, recombination cannot transfer the mean fitness of the selected parents completely from generation to generation. Thus the prediction of the fitness progress in one generation starts with $R(t) = h{\cdot}S(t)$, where h denotes the rate of *realized heredity* $(0 \leq h \leq 1)$. It can be seen from the figure that the increase of the average fitness in generation $t+1$ leads to a decrease of the fitness variance in the population at the same time. This fact is well known from breeding experience. As as sketched below, it can be proved under standard assumptions made for a GA.

For simplicity Mühlenbein assumes a simple fitness function known as the *one-max function.* In this problem, the fitness of a string is equated with the number of bits set to 1. Denoting the variance of fitness in generation t as $\sigma^2(t)$, the following equation holds if proportionate selection is used

$$S(t) = \frac{\sigma^2(t)}{\overline{f}_t}. \tag{3.3}$$

This analytical result verifies the above observation and allows us to predict the response to selection by $R(t) = h \cdot \sigma^2(t)/\overline{f}_t$ for a single generation[2]. Several conclusions can be drawn from the analysis.

[2]For $h = 1$ this equation corresponds to the fundamental theorem of population genetics, cf. Hofbauer and Sigmund (1984).

We only mention one important implication of the above model. Reaching a state where every bit has its optimal value is intended for the simulated evolution. If this aim is reached, the GA is said to have *converged* to optimality. However, equation (3.3) implicates that an infinite population cannot converge to optimality if proportionate selection is used. Although this claim seems rather theoretical, it is highly discouraging since it applies already to fitness functions being as simple as the one-max problem.

As a remedy Mühlenbein and Schlierkamp-Voosen (1993) propose to use *truncation selection*. Here only the $\tau\%$ best strings are selected as parents. Usually τ is chosen in between of 10% and 50%. This range is adopted from the artificial selection process carried out by animal breeder. In terms of the time needed for convergence, truncation selection has shown to be more effective for optimization than proportionate selection.

4. SUMMARY

In this chapter we have presented the fundamental ideas of EAs and we have outlined them in detail in the context of GAs. Actually, much of the algorithmic templates presented and parts of the background theory as well can be transferred to other members of the EA family.

Unfortunately, the current state of analytical models neither yield sufficient understanding why GAs work well in many applications nor provide practical guidelines for GA design. What basically turned out by theory is that the search progress of GAs strongly depends on the balance of the fitness variance in a population and the intensity of selection. This finding can be confirmed by computational experiments easily.

It is already clear so far that evolutionary computation cannot yield methods which are competitive or even superior regarding every problem of interest. The scope of problems which particularly address the potentials of EAs are those which are either intractable or prohibitively time consuming to be solved by other algorithms. In order to assess if it is worthwhile to tackle a certain problem with evolutionary computation methods, one must take a close look on either the properties of the problem or on the properties of the algorithm. We follow both directions in the subsequent chapter.

Chapter 4

ADAPTATION TO STATIC ENVIRONMENTS

The idea underlying adaptation in biology is pretty simple as it rouses the notion of walking and climbing in a fitness landscape. The evoked image is powerful since it allows us to consider search as a process which only relies on the properties of the particular landscape observed. Thinking in terms of natural landscapes such as *peaks*, *valleys* and so on can enhance insight into the difficulty of adaptation, but it can also give rise to new strategies for exploring complex search spaces.

The chapter investigates the capabilities of adaptation for population-based search techniques developed in evolutionary computation. The requirements of successful adaptation are approached from two directions. In the first section we take a close look at the adaptive behavior of GAs. In next section we investigate properties of problem environments in reference to a fitness landscape. In order to gain insight into the potentials of evolutionary search we assume a static environment throughout this chapter, i.e. we consider a fitness landscape as fixed in structure. The notion of a landscape, however, is abstract enough to encompass a dynamic environment as well. This will be investigated later on.

The last section presents the basic principles of evolutionary search in the context of other local search techniques. In combinatorial optimization EAs can be an alternative for algorithms like Simulated Annealing or Tabu Search. Since the landscape model is restricted neither to specific problems nor to the viewpoint of specific methods it provides a framework suitable to access general concepts of local search. By predicting the difficulty of local search with respect to the structure of a fitness landscape we finally attain a cautious assessment of EAs and the scope of problems they are supposed to be applied to.

1. CONVERGENCE IN EVOLUTIONARY ALGORITHMS

The dynamic behavior of EAs must be taken as the most crucial point regarding their performance. As outlined in the previous chapter, the dynamic of evolutionary search mainly depends on the intensity of selection. In order to drive a population of individuals towards promising regions of a search space the gathered information is continuously condensed. If this process stagnates, the algorithm is said to have converged. If a population does not converge, it can hardly adapt to a given fitness function. On the other hand, a converged population cannot yield further adaptation. For this reason convergence and adaptation are closely tied processes and their control is an important need in order to make an algorithm work.

This section is divided into three parts. First we provide a method for measuring the state of convergence in a population of binary strings and then we review recent approaches to control the convergence process. Finally a distributed control model of local recombination is presented which enables the individuals of a population to react flexible on their state of convergence.

1.1 MEASURING CONVERGENCE

Often a particular GA performs well for relatively small optimization problems but it leads to poor results if it is applied to larger instances of the same problem type. In order to achieve a more thorough search we may react by increasing the population size of the GA. But, as mentioned before, there is only little trade-off for using a larger population. The attempt to improve search by enlarging the population at sufficient size is usually computational prohibitive. In natural evolution time is of no concern and adaptation can rely on evolving huge populations. In simulated evolution this is different and therefore research has spent effort to gain more insight into the requirements for adapting towards near-optimal solutions.

As we have learned from population genetics, the average fitness of a finite population tends to stabilize from generation to generation. The reason that the individuals show increasingly similar fitness results from a similarity in terms of their information encoded. In other words, the behavior of GAs provides a dynamic which can result in stagnation at the information processing level which is called *convergence*. A qualitative description of GA convergence is given as follows: in order to gain further progress, above average fit solutions are selected at high rates for recombination. If these partially adapted strings become similar, the

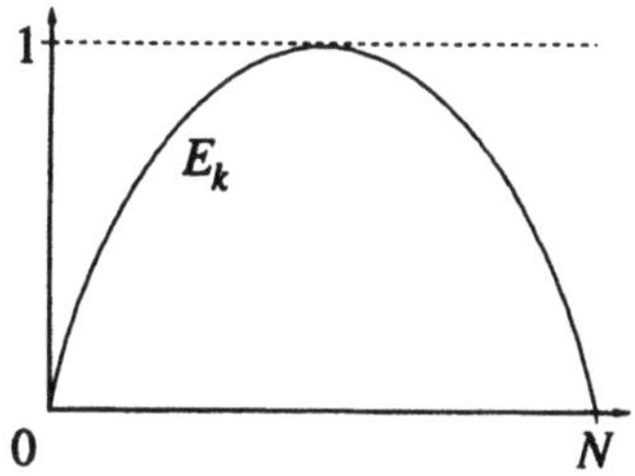

Figure 4.1. The entropy E_k at bit position k as a function of bit frequencies ($0 \leq \omega_k \leq N$), observed in a population of μ individuals.

diversity of the population decreases. In consequence, the ability of the reproduction mechanism to discover new strings becomes smaller and finally the strings are too similar to allow further substantial progress.

In order to assess GA convergence in quantitative terms a measure is needed which computes the diversity of the binary strings within a population. Such a measure is derived from the general concept of *entropy*[1] which allows to quantify the amount of disorder in a system. Let $\omega_k(1)$ denote the frequency at which the 1-bit occurs at the k-th bit position in the μ binary strings of a population. The frequency of the 0-bit at bit position k is given by $\omega_k(0) = \mu - \omega_k(1)$. The entropy E_k in the population concerning this bit position is calculated by

$$E_k = \frac{-1}{\log 2} \sum_{b=0,1} \frac{\omega_k(b)}{\mu} \log\left(\frac{\omega_k(b)}{\mu}\right). \tag{4.1}$$

Notice that E_k is close to one if both bit values occur with similar frequency at bit position k. If this balance diminishes at that bit position, i.e. either the 0-bit or the 1-bit begins to dominate the position, the entropy falls towards zero. This property is verified by the curve of the entropy function shown in Fig. 4.1. Accordingly, the total entropy of a population of binary strings of length n is given by $E = \frac{1}{n}\sum_{k=1}^{n} E_k$. In this way the entropy provides a general measure of the disorder at the information processing level of search algorithms, compare Dowsland (1993). As argued before, the amount of disorder in a GA population is responsible for its ability to effect substantial changes. Ideally, while the GA progresses the entropy declines from $E=1$ for the initial population to $E=0$ where the population has converged to a global optimum. Independent of whether a GA actually converges to optimality, its behavior is likely to approximate the sketched state change of entropy.

[1]The concept origins from physics where it is used to describe the molecular state-changes of a homogeneous mass in a thermal process.

Now we can estimate the adaptability of a GA at a certain state by means of the entropy. If the entropy of the population decreases, the GA increases its potentials to adapt to the environment and vice versa. If, however, the entropy falls below a predefined level (close to zero), the algorithm is expected to have exhausted its potentials for adaptation. Note the practical importance of this consideration: the entropy measure E provides a flexible criterion for terminating a GA.

1.2 CONVERGENCE CONTROL APPROACHES

The event occuring when a population of strings attains a low level of entropy at all bit positions at an inferior level of solution quality is referred to as *premature convergence*. In order to avoid premature convergence some approaches have been proposed to maintain the diversity of strings in the population over longer periods. The basic idea is to take care that recombination predominantly works on individuals which differ to a considerable degree, which is also called *incest prevention*. The simple rationale of this idea is to decrease the probability that a sub-optimal solution can dominate a population within a few generations.

Goldberg and Richardson (1987) first take the sharing of individuals into detailed consideration. In their approach the fitness of an individual deteriorates if the population contains many other individuals which are nearby in the problem space. By using the proximity of solutions as an indicator for similarity in the binary search space, the mechanism limits the uncontrolled spread of resembling strings within a population.

Related approaches are based on measuring the similarity of individuals directly in the search space. In boolean spaces a common measure of distinction is provided by the *Hamming distance*. Given two binary strings s_1 and s_2 of length n, the metric HD compares s_1 and s_2 by counting the number of differing bit positions, i.e $0 \leq HD(s_1, s_2) \leq n$ holds. Eshelman and Schaffer (1991) propose a threshold control for the crossover of binary strings. If the Hamming distance of parental strings falls below the threshold, the recombination step is prevented. Whenever the population cannot adapt anymore, the threshold is decremented in the next generation. A survey of these techniques is given by Davidor (1991).

Using the sketched approaches causes a retardation of GA convergence which can lead to considerable improvements of the performance. Still, computational experiences show that the progress is usually not substantial enough to enhance the scalability of a GA significantly.

Other approaches to control GA convergence base on distributed populations as introduced in Chap. 3(2.1). If a population is structured

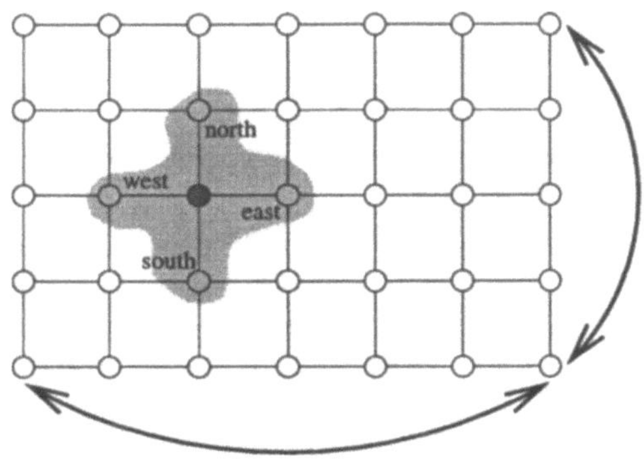

Figure 4.2. Spatial population residing on a toroidal connected grid.

either in "demes" or "islands" the information flow between individuals is restricted. This feature is expected to preserve the diversity of strings within useful bounds. In distributed populations selection and reproduction are done locally. According to Gorges-Schleuter (1992) this is referred to as *local recombination.*

1.3 A CONTROL MODEL OF LOCAL RECOMBINATION

In spatially distributed populations, recombination relies on a small number of potentially selected parental strings. Here global premature convergence is alleviated at the expense of incest in the localities of individuals. A remedy is suggested by the idea that an individual is able to change the behavior as a function of its own environmental conditions. In the initial phase of simulated evolution incest hardly occurs and therefore individuals can rely on the ordinary selection-recombination scheme. In later phases, when the population is threatened with convergence, an individual should avoid pointless crossover attempts with very similar partners of its locality. Performing a mutation instead is often the better strategy because it increases the diversity of strings which in turn increases the probability of useful crossover operations in the following generations.

In the population model proposed by Mattfeld et al. (1994) individuals reside on a toroidal connected grid. In each generation the individuals can either perform an action (i.e. a crossover or a mutation) or they can behave passively which is referred to as *sleeping.* An action produces a single offspring which either shows an improved or a weaker fitness with respect to the parent. The offspring replaces its parent if its fitness is not significantly below the fitness of the parent. Hence, a slight deterioration

Table 4.1. A control structure for local recombination.

condition	*action*	*outcome*	*threshold adjustment*
superior individual	sleeping	—	$T =: \max(1, T + 0.02)$
$\phi \leq T$, *HD* is ok.	crossover	improvement	$T := \max(1, T + 0.05)$
		accept only	$T := T$
		reject	$T := \max(0, T - 0.02)$
$\phi \leq T$, *HD* is too low	sleeping	—	$T := \max(0, T - 0.20)$
$\phi > T$	mutation	improvement	$T := \max(1, T + 0.15)$
		accept only	$T := \max(1, T + 0.05)$
		reject	$T := T$

of the fitness is accepted, otherwise, however, the offspring is rejected. An individual that performs no active operation automatically avoids replacement by offspring.

The control model of local recombination works as follows. First an individual checks the similarity of fitness within its locality. According to the compass rose it compares its fitness with the fitness of its four neighbors on the grid, see Fig. 4.2. If the fitness is superior to all four neighbors, the individual performs no active operation in order to prevent replacement. Otherwise the behavior of the individual is drawn probabilistically by a uniform random variable $\phi \in [0, 1]$. If ϕ exceeds a threshold T the individual calls the mutation operator and otherwise it calls the crossover operator. Decreasing T increases the probability of mutations. Initially the threshold is set to $T = 1$ which enforces crossover. Before the crossover operation is performed, a partner string is selected on the grid by probabilistic ranking from the least-fit to the most-fit neighbor with $p_x \in \{0.1, 0.2, 0.3, 0.4\}$ respectively. Before crossover is actually performed, both parents are checked for similarity by means of the Hamming distance. If the both individuals are almost identical, crossover is omitted.

The control structure of this recombination model is shown in Tab. 4.1. Altogether eight responses are distinguished which are tied to different reinforcements of the threshold. The adjustment of T follows plausible rules in order to adapt the behavior of each single individual towards the environment of its actual neighborhood. The control reacts on incest with a strong decrease of T which in turn increases the frequency of mutations in order to introduce new information in further generations. If a mutation yields an improved individual, the threshold is increased which in turn increases the probability of crossover again.

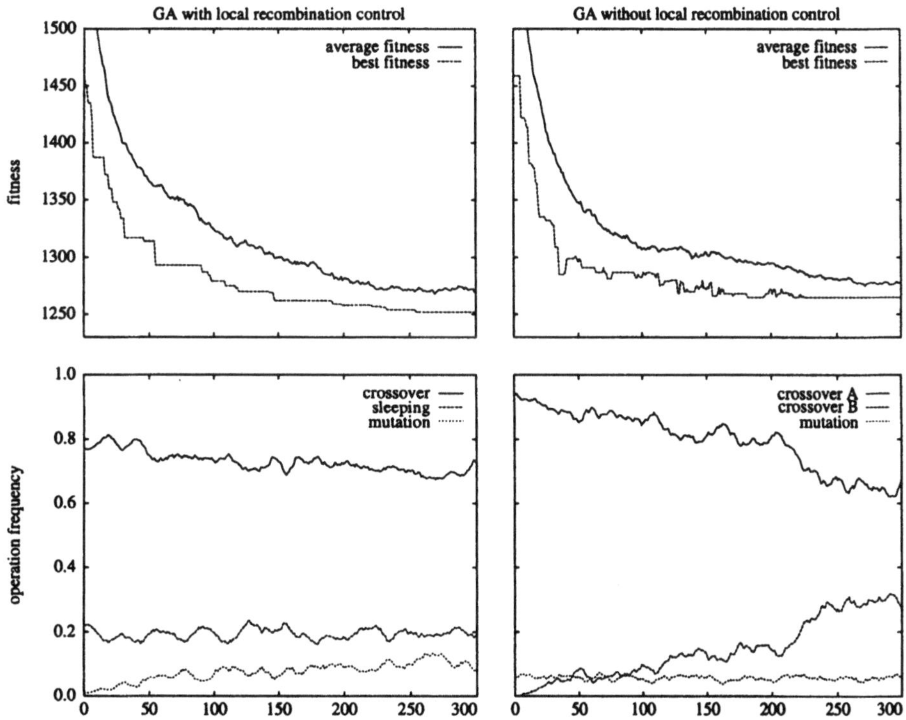

Figure 4.3. Influence of the control model on the population dynamic.

The approach has been tested by Mattfeld (1996) for solving job shop scheduling problems. In a comprehensive study a suite of 162 benchmarks of this problem class is solved by a GA. The local-recombination control reduces the observed average relative error by 12.5%. Due to the concept of "sleeping individuals", the runtime of the GA is shortened considerably at the same time. In the problem class under consideration the objective function has to be minimized. Thus the fitness function is directly derived from the objective function such that the fitness is minimized by the GA.

Fig. 4.3 compares the population dynamic effected by the local recombination model with the performance of an ordinary GA. Both algorithms maintain a distributed population of 100 individuals which are located on a 10×10 toroidal grid. The same genetic operators are used. In the ordinary GA the crossover rate is set to 1.0 and the mutation rate is set to 0.05.

The upper diagrams show the progress made by both variants over 300 generations in terms of the best and average fitness. It can be seen clearly that the local-recombination control slows down the convergence in early generations and retains average fitness longer. The ordinary

GA has converged prematurely after 200 generations whereas its rival generates improvements beyond this point. The lower pictures report on the frequencies of calls to the genetic operators over time. Due to the fixed mutation rate the mutation frequency is almost constant for the ordinary GA. Calls of the crossover operator are divided into two classes. A crossover falls into class A if the parental strings have a Hamming distance larger than 1% of the string length. This case is shown in the upper curve. It can be seen that the frequency of useful crossover operations decreases approximately linear in time. To the contrary, the increasing lower curve documents the frequency of pointless crossover operations. This class B represents the incest rate in the population.

The corresponding picture for the GA using the local-recombination control indicates a totally different dynamic. The crossover frequency decreases slowly with a slight increase of mutations. The frequency of the sleeping operation is fixed at about 20%. During the initial phase of the GA sleeping is mainly caused by superior fitness, whereas it is mainly triggered by incest trials in later generations. Amazingly, adding both cases leads to a nearly constant frequency of sleeping during the whole GA run. Notice that the sleeping process reduces the runtime considerably if the evaluation of individuals is time consuming.

The above model steps in a crucial aspect of GA design as it essentially renders the question of parameter tuning already discussed in Chap. 3(2.7). According to the classification of Angeline (1995) the approach can be referred to as a *self-adaptive evolutionary computation* because it enables the algorithm to change its behavior. Angeline distinguishes between population-level adaptive methods which adjust the parameters of an EA that are global to the entire population (including the population size) and individual-level adaptive methods. In this classification the model described above falls in the latter category because it determines how a particular individual is affected by the mutation operator in order to prevent a stagnation of search. However, it is by no means clear that increasing the mutation rate is always a promising strategy. To decide this question requires to investigate the search environment.

2. STRUCTURE OF SEARCH SPACES

This section deals with the famous metaphor used for accessing the structural properties of discrete search spaces. First we introduce the notion of a fitness landscape and then we review an abstract model for classifying the ruggedness of such landscapes. Finally statistical methods for analyzing the structure of fitness landscapes are provided.

2.1 FITNESS LANDSCAPES

The structure of search spaces has received continuous attention in evolutionary computation[2]. In order to access the structure of a discrete search space we have to consider the mapping of fitness values over the *problem representation space*, i.e. the configured set of encoded solutions. An analog finding was made in evolutionary genetics already in the thirties. At this time, the biologist *Sewall Wright* introduced the paradigm of a space of possible genotypes which led to the notion of the *fitness landscape*. This theoretical approach regards adaptation as progressing through small changes by involving a local search in the space of possibilities[3]. A series of consecutive transitions via fitter mutants is referred to as an *adaptive walk* in the fitness landscape.

The notion of adaptive walks apparently corresponds to the principles of hill-climbing and iterative improvement algorithms as developed in Artificial Intelligence and Operations Research independently. The discrete spaces considered there have often a combinatorial structure very similar to a genotype space. In order to conceptualize heuristics in search and optimization we borrow the term fitness landscape from evolutionary genetics.

Let us view a problem representation space as a structured set of possible solutions to a problem. We derive a fitness function from the objective function of the problem such that the fitness function is either maximized or minimized and assumed to be non-negative. Now we can suppose the fitness of all solutions as forming a contiguous surface over the representation space which we refer to as the fitness landscape of the corresponding problem.

Formally a discrete search space S consists of a finite number of elements. The structure of S is imposed by a topology which defines for each element $s \in S$ a set of *neighboring solutions* $\mathcal{N}(s) \subset S$. If arbitrary elements of S can be transformed into each other by a sequence of *neighborhood moves*, $\mathcal{N}$ is called a connected neighborhood. In this case the mapping

$$f : S \rightarrow \mathbb{R}, \qquad s \mapsto f(s) \tag{4.2}$$

defines the fitness landscape of a function f over S with respect to the definition of $\mathcal{N}$. In other words, a static fitness landscape refers to a

[2] see e.g. Manderick et al. (1991); Mathias Whitley (1992); Inayoshi et al. (1994); Jones and Forrest (1995); Horn and Goldberg (1995).

[3] For an outline of its utility in biology see Maynard Smith (1989).

mapping of a finite graph into the real valued numbers[4]. The vertices of the graph represent the elements of S and the connecting edges of vertices represent the neighborhood relations $\mathcal{N}(s)$ for all $s \in S$.

The notion of the fitness landscape leads to a model of the search space consisting of peaks, valleys, plateaus and the like. By interpreting adaptation as a process of guiding a local search towards the regions of high fitness in the landscape, these "visible" characteristics of landscapes promise some intuition of the intractability of a problem. At first glance the ruggedness of a landscape is accessed by answering the following questions:

1. How many peaks (local optima) exist in the landscape?
2. How smooth is the surface of the landscape?
3. Do regions of above average altitude exist in the landscape?

In the following we are going to transform the above questions into methods for classifying the difficulty of adaptation. For this purpose we first describe an abstract approach for generating landscapes of variable dimensions with a tunable structure. This model has been suggested by Kauffman (1993) under the name *NK*-Model.

2.2 THE NK-MODEL

The model is stated in biological terms, hence we follow this lead. Recall that a chromosome is the biological counterpart of an element s of the problem representation space S. A chromosome consists of a fixed number of N *genes* which are taken from a finite set of gene values. The positions of genes in a chromosome are called *loci* and the gene values which can occur at a certain locus are called the *alleles* of that locus. In biology it is assumed that each gene of a chromosome contributes to the overall fitness of an organism. On the other hand it is well known that the fitness contribution of one allele often depends upon the allele occuring at some other locus. This phenomenon is called an *epistatic interaction* of two genes. Since the alleles at all N loci may mutually influence each other, it is often almost impossible to determine the contribution of a certain gene to the overall fitness.

In reference to a boolean search space we can think of chromosomes as fixed-length strings over a binary alphabet $\{0, 1\}$. For this representation space we can define a simple neighborhood structure by assigning each

[4]By this transition, a static fitness landscape corresponds to the concept of *heuristic search in a state-space*. The algorithms developed in the field of AI are typically described in terms of a search on a graph, compare e.g. Kopfer (1989).

$\pi_i\varepsilon_i$	w_1	w_2	w_3
00	0.2	0.6	0.4
01	0.1	0.5	0.2
10	0.7	0.9	0.6
11	0.6	0.3	0.4

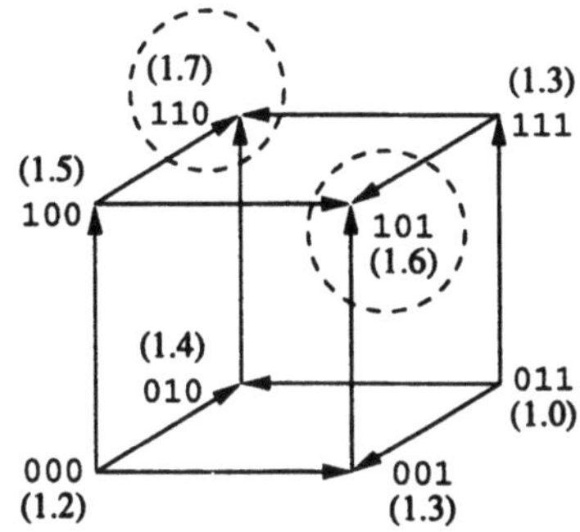

Figure 4.4. The fitness contribution of bits as a function of their locus and their epistatically interacting bit. The data in the table determines the overall fitness of 2^3 possible strings, like for instance $f(110) = w_1(11)+w_2(10)+w_3(01) = 0.6+0.9+0.2 = 1.7$. The corresponding fitness landscape, shown to the right, generates a three-dimensional cube.

string $s \in S$ all other elements of S that differ in exactly one bit position (*bit-flip neighborhood*). Let us consider binary strings of length N. We denote the average number of loci in a string which epistatically affect the fitness contributions of the alleles "1" and "0" at each certain locus by K. Obviously, K may range between zero and $N-1$. For $K=0$ the fitness contribution of the bit occuring at one locus is independent of the bits occuring at all other loci. In contrast, $K=N-1$ means that the fitness contribution of a single bit depends on all other $N-1$ bits. Thus N and K are the decisive parameters in the model which establish the family of *NK-landscapes*.

Example 4.1 We consider binary strings of length $N=3$. Let us assume that the fitness contribution of each bit is epistatically affected by one other bit position ($K=1$) as it is shown in Fig. 4.5.

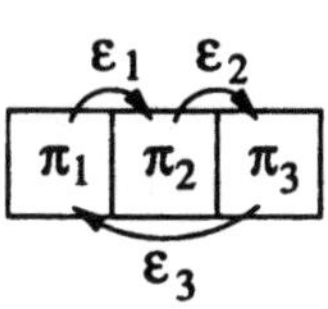

Figure 4.5. A simple model of epistasis with $N=3$ and $K=1$. The interaction is directed in clockwise rotation such that the last bit of a string affects the first bit.

We denote the bit value occuring at the i-th locus as π_i and its epistatically interacting bit value as ε_i. Consequently, the fitness contribution of the bit at the i-th locus depends upon its own value π_i and on ε_i. We model the epistatic interaction by assigning the four possible bit combinations "$\pi_i\varepsilon_i$" occuring at two adjacent loci a different fitness contribution. The fitness contribution w_i of the i-th locus is randomly drawn from the uniform distribution in $(0,1)$.

The corresponding fitness landscape is produced as a graph in Fig. 4.4. Notice that the edges of the cube correspond to the bit-flip neighborhood in the boolean space. The orientation of edges expresses the direction taken by adaptive walks along the landscape. By the two encircled solutions we recognize that a space of eight points can already possess more than one local optima. ■

The *NK*-model allows an abstract classification of the ruggedness of fitness landscapes. By varying the parameter K while leaving N constant, different structures of landscapes are generated:

- $K = 0$: The fitness contribution of a bit is independent of the other bits, i.e. optimal bits can be determined consecutively at each bit position. The landscape has a single optimum which is easy to find.
- $K = 1$: The fitness contribution of a bit depends on one other bit. The landscape has already many local optima, but adaptive walks are still very long because the average altitude increases in one region of the landscape.
- $K = N - 1$: The fitness contribution of a bit depends on all other bits. If only one bit is changed, the contribution of this bit changes but also the contribution of all other bits. As a consequence, there is a huge number of local optima and adaptive walks are very short.

If the influence of epistatic interaction increases by means of the tunable parameter K, the fitness of nearby solutions in the representation space correlates less and less. For $K = N - 1$ the fitness landscape is fully uncorrelated. In this way the *NK*-model classifies the ruggedness of fitness landscapes. If the number of epistatic interaction K remains small while N increases, landscapes retain high accessible local optima. In reverse, if K increases while N remains constant, the corresponding landscape tends to become more multi-peaked. In the worst case the average fitness of local optima falls towards the mean fitness of the landscape. In this situation a systematic search is hardly possible.

The epistatic interaction observed for a combinatorial optimization problem expresses the number of conflicting constraints involved. Notice, that the constraints of a problem do not necessarily conflict in the search space considered. In this case the epistatic interaction is low because partial solutions can be developed independently of each other to some extent. On the other hand, if the number of conflicting constraints is high, the epistatic interaction increases and partial solutions often result in incompatibility. The degree of epistatic interactions is not fixed for most combinatorial problem classes. As a consequence the corresponding landscapes cannot be specified in terms of the *NK*-model anymore.

2.3 STATISTICAL MEASURES

In the abstract terms provided by the *NK*-model the estimate of problem difficulty ranges from "easy" to "almost intractable". By emphasizing the structure of search spaces more generally we hope to get insight into what makes fitness landscapes difficult to adapt to. Therefore empirical measures are needed. This section proposes such measures in reference to the three questions raised on page 70.

THE MODALITY OF LANDSCAPES

The question addresses the number of peaks in a landscape by assuming that a low-modal landscape is more easy to search than a highly multi-peaked one. A first insight into the modality of a landscape results from the *dimensionality* of the underlying representation space. The dimensionality of a point s is defined by the number of other points that can reach s by a single neighborhood move. In other words the dimensionality is given by the cardinality of $\mathcal{N}(s)$. Unfortunately, the dimensionality of a problem representation space can differ at various points. Therefore the dimensionality of a landscape is defined as the average number of neighbors to each point.

For simplicity we concentrate for the moment on the boolean hypercube of dimensionality N. Here the possible solutions are represented by all 2^N binary strings of length N. The probability that any binary string is a local optimum is just the probability that it has higher fitness than any of its N bit-flip neighbors. For uncorrelated landscapes ($K = N - 1$) this probability is $1/(N+1)$. Consequently the modality M of the landscape is $2^N/(N+1)$. For an uncorrelated search space S of dimensionality D_S and cardinality C_S the modality is estimated by

$$M = \frac{C_S}{D_S + 1}. \tag{4.3}$$

In case of $K < N - 1$ the formula yields an upper bound for the number of local optima. It shows that the modality of a landscape may increase almost as rapidly as the size of the search space. Moreover we see that the modality reduces if the dimensionality of the space increases.

THE SMOOTHNESS OF LANDSCAPES

If the modality of a landscape is unknown, the question of its *smoothness* may stand proxy. In order to access the smoothness of landscapes empirically we focus on the length of adaptive walks. The points in the landscape which can reach a certain local optimum via adaptive walks establish the *basin of attraction* of this peak. The extension of a basin determines the force of attraction of the corresponding local optimum.

It is obvious that the larger the basins of attraction are, the *smoother* a fitness landscape is.

A clue to the smoothness of a landscape results from measuring the extension of its basins of attraction . An approximation is obtained from the average length of a sufficiently large number of adaptive walks starting from arbitrary points in the search space. Adaptive walks can be carried out by using any iterative improvement algorithm available (see Sect. 3.1). In case of a binary representation space the length of an adaptive walk is measured by the Hamming distance observed between the starting point and the reached local optimum. We denote the average length of adaptive walks as $\overline{r}_{adapt}$. The extension of the basins of attraction alone, however, is not expressive without reference to the overall extension of the landscape. Therefore we measure the distance $\overline{r}_{rand}$ between any of a large number of randomly generated points and an arbitrarily chosen reference point. Finally we calculate the *expected force of attraction* Λ of the local optima in a landscape by

$$\Lambda = \frac{\overline{r}_{adapt}}{\overline{r}_{rand}}. \tag{4.4}$$

It is defined as the percentage of the basins of attraction compared to the overall extension of the surrounding space. For landscapes with $K = 0$ we obtain $\Lambda = 1$. To verify this we just have to choose the only optimum existing as the reference point. In the other extreme we face a fully uncorrelated landscape with $K = N - 1$. According to Kauffman (1993) the average length of adaptive walks is estimated by $\overline{r}_{adapt} \approx \ln(N-1)$. Since $\overline{r}_{rand} = N/2$ holds in the N-dimensional hypercube we obtain $\Lambda \approx 2 \cdot \log(N-1)/N$ which converges to zero for large values of N. Thus Λ ranges in between zero and one representing the relative attraction of local optima in a fitness landscape. We use the measure for estimating the difficulty of adaptation: a large Λ value indicates comparable problem easiness whereas we expect to meet challenging problems if Λ is in the range of a few percent only.

Another way to access the smoothness of landscapes has been proposed by Weinberger (1990). This measure results from an unbiased walk on the fitness landscape which produces a series of samples of the fitness differential observed between neighbored points. An unbiased walk of length l results in a sequence of fitness values $f(t)$ $(1 \leq t \leq l)$ which can be interpreted as a time series of l lags. For statistical analysis it is presupposed that the landscape is *isotropic*. This implies that the autocorrelation of the time series does not depend on the particular samples taken on the walk. The autocorrelation of the time series regarding an interval of length $h < l$ is calculated by

$$\rho(h) = \frac{\frac{1}{l-h}\sum_{t=1}^{l-h}(f(t)-\overline{f})(f(t+h)-\overline{f})}{\frac{1}{l}\sum_{t=1}^{l}(f(t)-\overline{f})^2}. \tag{4.5}$$

The *correlation length* h^* is defined as the distance h for which correlation can be still observed, i.e. $h^* = \arg(\rho(h) = 0.5)$. The correlation length depends on the definition of neighborhood in the representation space and on the particulars of the problem instance considered. For *NK*-landscapes the correlation length is estimated by Weinberger with $h^* = -1/\ln(1-(K+1)/N)$. Computational results of Manderick et al. (1991) confirm this approximation.

The measurement of the correlation length can be used to compare different neighborhood operators by employing them for a walk on the landscape of a particular problem instance. Alternatively different problem instances of the same class and size can be compared. Assume that the correlation vanishes significantly slower for one instance while walking its fitness landscape, than for another. The longer correlation length indicates a smoother surface of the former landscape.

THE VARYING ALTITUDE OF LANDSCAPES

With respect to the landscape metaphor the term *massif* stands for a region of above average fitness in the landscape where a global optimum might be located. If such a massif exists, adaptive walks are attracted by the massif aside from the immediate attraction of local optima. In order to examine the existence of a massif we have to inspect if solutions of above average fitness are located close to each other in some region of the landscape. By again concentrating on local optima we already have representatives of above average fit solutions at hand. What remains to do is to check whether these local optima are located nearby each other.

In order to investigate the similarity of local optima the following procedure is used[5]. In a first step a sufficiently large pool of solutions is generated at random. We refer to this pool as *rand-pool*. Next adaptive walks are carried out, starting from each element in the rand-pool, and leading to a number of corresponding local optima. These resulting solutions are stored in a second pool, called *adapt-pool*. Finally the similarity of solutions in both pools is calculated and compared.

[5] cf. Taillard (1995).

A way to calculate the similarity of solutions is shown by the entropy measure, introduced in Sect. 1.1. Recall that a pool of very similar solutions has an entropy close to zero and a pool of uniformly distributed solutions has an entropy tending to one. Since it is not guaranteed that the possible solutions of a constrained combinatorial problem are uniformly produced by random generation (i.e. E_{rand} may be lower than one), we compare the entropy E_{adapt} in the adapt-pool with the entropy E_{rand} in the rand-pool. Certainly, E_{adapt} will be smaller than E_{rand} because otherwise adaptation has increased the entropy what is impossible for theoretical reasons. Hence, the *change of entropy* which is effected by adaptation ranges between zero and one.

$$\Delta E = E_{rand} - E_{adapt}. \tag{4.6}$$

This measure has a direct interpretation regarding the existence of a massif in the landscape. If ΔE is close to zero above average fit solutions can be found all over the landscape. In contrast, a large ΔE indicates that above average fit solutions are more frequent in a certain region of the landscape. Recall that the distance of points in the search space represents their similarity. Hence the existence of a massif simply indicates that above average solutions share a considerable amount of information.

In order to determine the change of entropy for a given landscape the entropy formula (4.1) is applied. If, however, solutions are not represented as binary strings, the formula needs a problem-specific transfer[6].

To summarize, we have presented different statistical measures for an empirical classification of the structure of fitness landscapes. A further approach of Jones and Forrest (1995), known as *Fitness-Distance Correlation*, is not presented here. Applying this measure requires to know the global optimum of a problem instance which will be given in very rare cases only.

In spite of the variety of statistical methods proposed to assess fitness landscapes, it strikes that only few problems of practical interest have been analyzed so far. Next to a bulk of research on local search algorithms and a large number of well studied combinatorial optimization problems, there is just the beginning of a literature concentrating on the structure of landscapes induced by such problems[7].

[6]For some problem classes the transfer is shown, e.g. for sequencing problems, assignment problems, and for scheduling problems, see Grefenstette (1987); Fleurent and Ferland (1994) and Mattfeld (1996), respectively.

[7]It includes graph partitioning and shop floor scheduling problems. The interested reader is referred to Stadler and Happel (1992); Inayoshi et al. (1994) for the former and to Bierwirth (1993) and Mattfeld et al. (1999) for the latter ones.

2.4 THE LANDSCAPE OF THE TSP

Among the *NP*-hard optimization problems the famous *Traveling Salesman Problem* (TSP) is certainly the most intensive one studied by academic research. Therefore it cannot surprise that fitness landscapes of this problem are comparably well-known. In this section we shortly review the major results concerning the structure of TSP landscapes.

The objective in the TSP is to find a route of minimal cost which starts at a home city a, visits some other cities b, c, d, etc., and ends at a again. The natural representation of the TSP is to encode the possible routes by fixed length strings containing all n cities involved such that the first position refers to the home city a. The order of cities in the string expresses the sequence they are visited. Several definitions of neighborhood are common, e.g. the pairwise-exchange of cities, its generalized k-exchange variant, inversions of a sublist, and the more sophisticated neighborhood of Lin and Kerninghan (1973). But let us for the moment concentrate on the most basic neighborhood $\mathcal{N}_B$ of the TSP which simply inverts adjacent cities of a route, Obviously, $\mathcal{N}_B$ is a subset of all other neighborhoods mentioned before.

Example 4.2 For simplicity we consider a small TSP consisting of $n = 4$ cities. Furthermore we assume that the TSP is asymmetric, i.e. the cost incurring if two cities are visited immediately after each other is not necessarily equal to the cost incurring if these cities are visited just in inverse direction. For a general n-city instance of this type of TSP there is a total of $(n-1)!$ different routes. Since every route has $n-2$ neighbors with respect to $\mathcal{N}_B$, the dimensionality of the search space is $n - 2$.

Another way of representing solutions of the TSP is to encode the precedence relations among cities as they can appear in a route, see the left table in Fig. 4.6. In this representation a bit position i refers to the relation $\text{prec}_i(x, y)$ of two specific cities x and y, where $\text{prec}_i(x, y) = 1$ means to visit city x (not necessarily immediately) before city y. Representing an n-city instance requires a total of $N = (n-2)(n-1)/2$ precedence relations. For instance string $s = \texttt{110}$ means to visit city b before c $(i = 1)$ and before d $(i = 2)$ and to visit city c after d $(i = 3)$. It can be seen from the table that two binary strings cannot be transferred into a feasible route. For this reason local search algorithms for the TSP rarely use the binary representation.

Nevertheless, the difficulty of the TSP is well demonstrated by reflecting its landscape on the basis of the boolean geometry. Supposed that infeasible strings are swapped into feasible ones by a minimal number of bit-flips, the possible routes are still produced with identical probability. As shown by the dashed cube-edges in Fig. 4.6, each infeasible string

s	*route*
111	*abcd*
110	*abdc*
101	-
100	*adbc*
011	*acbd*
010	-
001	*acdb*
000	*adcb*

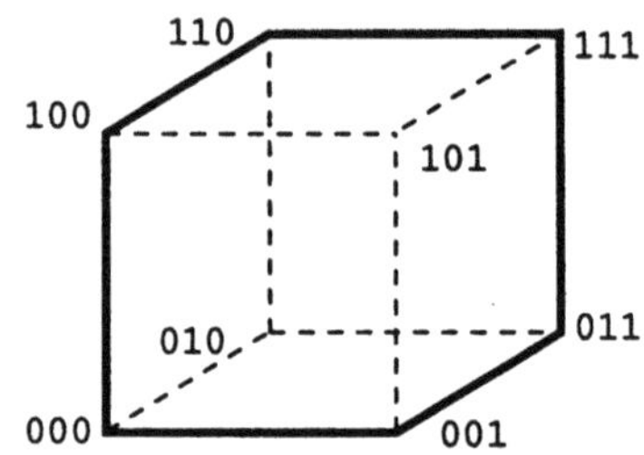

Figure 4.6. Binary vs. natural representation of routes for a small TSP. The bold edges of the cube show the structure of the representation space with respect to the basic TSP neighborhood. A feasible bit-flip corresponds to single $\mathcal{N}_B$-move.

can result in three different solutions. The bold edges indicate that the search space forms a ring where each route has exactly two neighbors. We verify from the figure that the size of the search space is smaller than the regular cube ($C_s = 6 < 2^3$). Since the dimensionality is smaller too ($D_s = 2 < 3$) we obtain the same estimated modality ($M = 2$) for both spaces from equation (4.3). In the extreme, already every $(n-1)$-th route could be expected to be a local optimum. This ratio increases less drastically in the surrounding N-dimensional hypercube where at most every $(N+1)$-th point is a peak. ∎

Thus far we have an idea of the modality of landscapes as induced by ordering problems. The combinatorial structure reduces the dimensionality of the representation space which increases the modality of the landscape again. Since the modality of the TSP is generally very high, we ask for the smoothness of its landscapes. TSP landscapes fulfill the condition of statistical isotropy, compare Weinberger (1990). Therefore the correlation length of unbiased walks on the landscape produces a reasonable measure of smoothness. For symmetric instances with uniformly distributed distances the correlation length can be estimated with respect to the neighborhood used. For large numbers of cities, Stadler and Schnabl (1992) approximate $h^* \approx n/4$ and $h^* \approx n/2$ for the pairwise exchange neighborhood and the sublist inversion neighborhood respectively. The observation that the TSP shows twice the correlation for inversions than for pairwise exchanges has been verified experimentally by Manderick et al. (1991). However, a linear increase of correlation with respect to n has good prospects regarding adaptation in both cases.

Finally we take a global look at the landscapes of symmetric TSP instances. Due to the property of statistical isotropy, random solutions are uniformly distributed, i.e. $E_{rand} = 1$ holds. For 2-optimal routes an entropy of $E_{adapt} = 0.32$ is reported by Taillard (1995). The strong change of entropy $\Delta E = 0.68$ confirms a previous result of Kirkpatrick and Toulouse (1985), saying that 2-optimal routes share two-third of their edges on average. This proofs the existence of a strong massif in the landscapes of the TSP where the best local optima are located very close to each other.

For a final remark on the TSP we mention that it applies to the *NK*-model with $K = 2$. Note that the cost contribution of each city solely depends on those two cities which precede and succeed this city in a route. From the above analysis we may conjecture favorable properties of TSP landscapes regarding the success of adaptation. This assessment is verified by the abstract model of Kauffman (1993).

3. ADAPTATION AND LOCAL SEARCH

Unlike research in evolutionary computation, research in combinatorial optimization has paid little attention to the structural properties of fitness landscapes. This wonders since current research is often unable to explain adequately why or when a certain local-search based heuristic is likely to succeed or fail. Computational studies have shown that none of the diverse approaches cuts out all of its rivals in solving different problems. Moreover it turned out that even when applied to different instances of the same problem class often no algorithm is superior.

In order to investigate local-search based algorithms generally in the context of adaptation, the section first transfers the notion of adaptive walks to the common terms used in local search. Then we consider a local search template which captures GAs among other local search algorithms. Finally we discuss the suitability of different algorithms for coping with the particular properties of fitness landscapes.

3.1 ITERATIVE IMPROVEMENT ALGORITHMS

The principle of local search is to perturb existing solutions slightly in order to gain improvements. The most simple variant of local search follows the idea to make iterative improvements. Given a combinatorial optimization problem and a suitable neighborhood definition, an *iterative improvement algorithm* starts from an arbitrary solution and then continually searches the neighborhood of the current solution for a solution of better quality. If such a solution is found, the algorithm replaces the current solution. Iterative improvement algorithms can apply either the *first-improvement*, in which the current solution is replaced by the first improving solution found, or the *best-improvement*, in which the current solution is replaced by the most improving solution respectively. The algorithm terminates if the current solution has no neighboring solution of better quality, i.e a local optimum has been found.

If we equate the terms "quality of a solution" and "fitness of a solution", the sketched principle corresponds almost completely to the notion of adaptive walks as developed in biology (see Sect. 2.1). By this replacement an iterative improvement algorithm performs a walk in the fitness landscape of the given problem instance where each step results in a fitter solution. Moreover, as adaptation is impeded in face of a rugged fitness landscape, an iterative improvement algorithm is also likely to fail. The algorithm can only succeed if its starting solution happens to be attracted by a local optimum of satisfactory fitness.

Note that algorithms for iterative improvements and EAs show apparent similarities. Both represent solutions in a vector scheme and both use syntactical search operators. In the simplest case a perturbation of a solution in the iterative improvement algorithm corresponds to a mutation in the EA. The search mechanisms engaged by iterative improvement algorithms form a subset of the mechanisms used by EAs. Thus one could argue that whenever an EA can be applied, basically an iterative improvement algorithm could do the job as well. There is, however, a serious objection addressing the efficiency of iterative improvement algorithms in general. If determining the fitness of candidate solutions is computational expensive or if estimating the success of perturbations efficiently is impossible, the time needed to verify local optimality is often not worth the effort spent. Therefore research concentrates on the construction of problem-specific neighborhoods which can be searched efficiently. Nevertheless, as long as suitable neighborhoods are unknown an EA is likely to work more efficient than simple local search.

Despite the fact that finding near-optimal solutions by iterative improvement algorithms is rather hopeless in multi-modal landscapes, the principle can still be utilized for practical needs. A lot of effort has been spent to make local searchers more powerful. The common idea of such approaches is to involve a meta-control which navigates search more properly than a first-improvement or a best-improvement strategy. Currently these meta-heuristics are intensively studied[8]. Since GAs are also involved in this literature we present them within a local search template. As a result, similarities between GAs and other popular meta-heuristics are emphasized.

[8]compare Pesch and Voß (1995) for a survey, and for comprehensive overviews Reeves (1993); Colorni et al. (1996), and Aarts and Lenstra (1997).

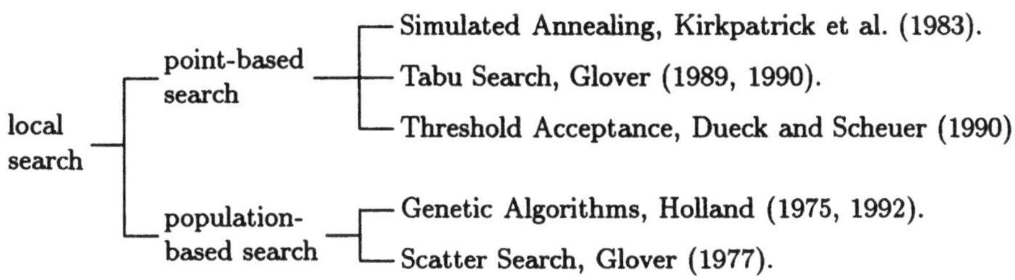

Figure 4.7. Classification of some algorithms according to item (i) of the template.

3.2 A LOCAL SEARCH TEMPLATE

In order to remedy the risk of getting stuck in poor local optima, various extensions of the basic principle of local search have been developed. For our purpose it is not necessary to consider these local search algorithms in detail. Instead we concentrate on a local search template adopted from Vaessens et al. (1992) which is general enough to capture most variants proposed. The template generalizes the iterative improvement algorithm with respect to

i. the number of the current solutions,

ii. the operators used for search, and

iii. the levels where the search may proceed.

Let us cover item (i) first. A local search algorithm which maintains a single current solution performs a so called *point-based search*. Such algorithms combine the iterative improvement algorithm with the opportunity to accept a neighboring solution with deteriorating fitness as the new current solution. In this way the algorithm can escape from the basin of attraction of already visited local optima in order to explore other parts of the search space. Prominent representatives of point-based search are *Simulated Annealing*, *Tabu Search*, and *Threshold Acceptance*, see Fig. 4.7. Another modification of the basic principle is to maintain more than a single candidate solution at the same time. In this so called *population-based search* we either can perform an independent parallel search or we can combine various existing solutions into new candidate solutions. Algorithms which perform population-based search include GAs among other EAs but they also include *Scatter Search*[9]. A gener-

[9] In Scatter Search new solutions are produced by linear combinations of an extracted subset of the current solutions. Although there are some similarities in Scatter Search and in EAs,

alized procedure which covers algorithms for point-based search as well as population-based search reads as follows:

1. A population of new solutions (containing one or more elements) is derived from the current population by using one or more *search operators.*

2. A *search guidance* picks promising solutions from both populations in order to produce the current population for the next iteration.

According to this framework, local search algorithms are distinguished by the search operators used for generating new candidate solutions and by the strategy used for guiding further search. The design of these operators leads us to item (ii) of the local search template.

Basically any search operator follows one of two simple rationales. A *perturbation operator* introduces new information by slightly altering one originator solution, expecting that the resulting nearby candidate shows a comparable fitness. A *recombination operator* recombines existing information by assuming that successful candidates can be assembled from pieces taken from various originator solutions.

The search guidance of an algorithm can determine the most improving candidates to replace the current population in the simplest case. In order to escape from local entrapments point-based search has to accept deteriorating fitness values by way of exception. But the rule to take improvements whenever possible remains predominant in these approaches. Therefore we refer to this control rationale as *guidance-by-improvement.* Guidance-by-improvement is still coupled with the tendency of getting stuck in local entrapments. The remedy proposed by population-based search is to maintain a population where the candidates compete for a placement in the population for the next iteration. This competition is performed on the basis of the fitness values observed in the population. Therefore we refer to the guidance rationale of population-based search as *guidance-by-selection.* Guidance-by-selection is limited to search situations where the fitness within the population differs significantly. Whenever the best fitness falls towards the mean fitness guidance-by-selection is likely to fail.

---

the approaches are sufficiently distinct and should not be mixed up with each other. As emphasized by Glover (1995), Scatter Search takes advantage from diverse search features which are not incorporated in traditional GAs, like for instance the strategic grouping of solutions or the simultaneous use of more than two parents to produce new solutions. Meanwhile, however, these and other ideas of Scatter Search have also influenced the development of GAs, see e.g. Eiben et al. (1994) or Mühlenbein and Schlierkamp-Voosen (1994).

3.3 INFERENCES FROM FITNESS LANDSCAPES

Thinking in the framework of fitness landscapes allows a straight interpretation of the search operators: a mutation (or perturbation) is viewed as a single move on the fitness landscape and a crossover is viewed as making a jump between the points involved in the operation.

If both operators are employed by an algorithm one may argue that local search must be seen as operating on at least a pair of landscapes. For this evidence Jones (1995) argues that GAs must be seen as operating on the *mutation landscape* and the *crossover landscape*. Since these landscapes will differ in their characteristics they have actually to be studied independently from each other. This has been subject to computational experiments carried out by Manderick et al. (1991). Their results confirm the hypothesis that the better an operator works, the higher its correlation length is. Therefore we confine to analyze one particular landscape of a problem, namely the landscape which is induced by the basic neighborhood of that problem. In doing so we claim that the difficulty of a combinatorial optimization problem manifests already from the perspective of the least perturbing operator available. For a binary problem representation space used by traditional GAs this demand calls the bit-flip operator. For an order representation space used by order-based GAs, a mutation typically results in a slightly larger modification of originator solutions. But, as shown in Example 4.2, the landscapes of ordering problems can still be assigned to a boolean space, where the bit-flip neighborhood applies.

Using the local search template we get reasonable insight into the potentials of local search operators for successfully adapting to the structure of a fitness landscape.

- For perturbation operators the smoothness of a landscape is indispensable for functioning properly. Otherwise, if the force of attraction is too low in the landscape, adaptive walks have hardly a chance to produce sufficiently large numbers of iterative improvements.

- For recombination operators the smoothness of a landscape is obviously of less importance. Here a condition of success is seen in the existence of a massif in the landscape. Since this indicates a partial similarity of local optima, assembling solutions from already existing pieces is promising.

Implications on a proper search navigation is recognized regarding the modality of the landscape.

- Guidance-by-improvement is impeded in face of a large number of local optima. Guidance-by-selection may work better for such spaces, because the search is acting more globally.
- Guidance-by-selection, however, is likely to fail if the landscape hardly shows a massif. Whenever the altitude of accessible local optima falls towards the mean fitness, the control looses the ability to navigate search properly.

By these assessments, point-based search and population-based search show different advantages and weaknesses regarding the structure of fitness landscapes. Based on the above considerations we may conjecture that a GA is a suitable search method if the landscape of a problem shows a strong massif. Since we have identified such a structure for the TSP it cannot surprise that GAs succeed for this problem[10]. In reverse, it can be conjectured that GAs are outperformed by point-based methods if the landscape of a problem lacks a massif but still shows correlation. Here a remedy might be to mix up perturbation operators and recombination operators with aspects taken from guidance-by-improvement and guidance-by-selection as well. A widespread representative of this class of algorithms is genetic local-search, see Chap. 3(2.6). Such approaches are addressed by item (iii) of the local search template.

4. SUMMARY

We have described easy to measure properties of static fitness landscapes which are helpful to predict the performance of EAs as well as other local-search algorithms. We have attained this method by relying on properties taken from the view of natural landscapes such as peaks, smoothness and mountain ranges.

One limitation of this approach is that it is by no means clear whether the convenience of the landscape metaphor scales up to problems with a huge number of dimensions. Another limitation concerns the statistical measures themselves. Although they show the relationship between a fitness landscape and the success of adaptation, they do not quantitatively predict this success. Nevertheless, verifying the measure of landscape properties within passable bounds appears advisable if not indispensable before entrusting a problem to an adaptive search method.

[10]see e.g. Freisleben and Merz (1996).

Chapter 5

ADAPTIVE AGENTS

Agent technology is one of the most exiting concepts studied in computer science at present time. The interdisciplinary research field aims to understand and to control the behavior of complex systems by synthesizing autonomously acting entities which are referred to as *agents*.

Basically an agent is a computational system which is able to flexibly solve contextual problems within a changing environment. Similar to a rule-based system, agents perform three functions, namely perception of the environment, derivation of actions from perceptions, and implementation of actions to effect the environment. Even so, agent technology has a different background than traditional AI as it focuses on the power of agent interaction and reactive behavior rather than on domain knowledge and inferences.

The capabilities of an agent can range from very limited to highly sophisticated. This wide taken understanding allows a variety of different techniques to be designated as agent approaches. A major branch of research concentrates on the desire to design agents capable of changing their behavior on the basis of previous experiences. These so called *learning* or *adaptive agents* profit from the use of methods which realize a learning-like process on a machine basis. Among other paradigms, evolutionary computation provides a class of such methods.

The particular features of EAs that satisfy the requirements of machine learning are discussed in three section. First we deal with the ability of EAs to sense, act, and communicate in a given environment. Applying EAs as one possible variant of machine learning is outlined afterwards. Based on these elements a framework for evolutionary adaptive agents is presented in the last section.

1. EVOLUTIONARY ALGORITHMS AS AGENTS

The fundamentals of agent technology are briefly introduced. Thereby we concentrate on the question to what extent agents differ from ordinary algorithms in general and in particular from EAs. By taking GAs into consideration we examine whether the characteristics of the method justifies it to call them agents.

1.1 AGENT DEFINITIONS AND PROPERTIES

The central requirement of a computational agent is to respond purposeful to the range of situations it be can confronted with. It is generally assumed that agents are *autonomous* to a certain extent, i.e. they must have control over their actions without necessarily receiving an external intervention. The most simple environment faced by an agent alternates between two different states. In this case an agent must possess the capability to switch like a thermostat. Such environments usually cause a stereotyped agent behavior. More complex environments change along several dimensions such that, from the viewpoint of the agent, the available resources vary or the measure of performance alters. Agents should react adaptable to such events which entails their ability to generate a versatile behavior.

Depending on the environment computational agents can take many different forms. Real world agents are usually designated as robots. In contrast, *software agents* are disembodied entities which are implemented on computer systems or in networks. Following Maes (1995), one can describe software agents as

> computational systems that inhabit some complex, dynamic environment, sense and act autonomously in this environment, and by doing so realize a set of goals or tasks for which they are designed.

At a first glance this definition does not seem not to allow to distinguish between ordinary algorithms and software agents. Let us take a look for instance at an EA which is employed for optimization. The algorithm certainly is a computational system which senses its environment by means of the data and the constraints involved in the problem under consideration. Moreover the algorithm pursues a goal which is to optimize a given objective function. Since the above definition restricts environments to be complex and dynamic it finally rules out EAs as agent, at least if the optimization problem is of static nature.

A more pervasive agent definition is due to Franklin and Graesser (1997). They designate an agent as

> a system situated within an environment that senses that environment and acts on it, over time, in pursuit of its own agenda and so as to effect what it senses in the future.

Most striking in this definition is the stress of time aspects. Returning to the above issue we have to admit that an EA does not meet the "over-time" condition. Even if the algorithm is run several times in sequence the output of one run will certainly not effect any future input, i.e. what the EA senses in later runs. Most ordinary programs which perform a certain static task are not software agents, whereas software agents are of course always programs.

Contrasting the tenor of this finding we argue in the following that EAs yet can be considered as computational agents if they operate in dynamic environments. So far we have ignored this feature of EAs because their use was limited to environments without temporal dimension. In even more generality Franklin and Graesser (1997) claim that a system which is not an agent can become an agent if a new dimension is inserted in the environment. In reverse, neglecting a single dimension of an environment may cause a system to be no longer an agent. As an example they consider a robot exploring a room with only visual sensors. If the light is switched off, the robot is simply a machine but no agent anymore.

In order to approach a global definition for computational agents it is helpful to list their essential properties. A sufficient understanding how we use the term "agent" in correspondence to EAs is disclosed by the following enumeration adopted from Franklin and Graesser (1997).

i. Agents are *reactive*, through sensing and acting in the environment they can respond to environmental requirements.

ii. Agents yet do not simply act in response to the environment, they behave *goal-oriented.*

iii. Agents are *permanent* entities. Within its life-span an agent is a continuously running process.

iv. Agents are *communicative*. They may negotiate to make agreements with each other, and they may support each other by exchanging knowledge.

Next to these definite properties of agents, an additional attribute is of particular interest for us[1].

[1] Further attributes are treated in literature, e.g. *mobility* and *personality* of agents. Since these properties mainly point to robot agents, they are outside of our scope.

(v) If a computational agent is able to change its behavior based on previous experience it is called an *adaptive agent* or *learning agent*[2].

The need to change behavior can be triggered by various conditions met in the agents environment. According to Hayes-Roth (1995) computational agents generally aim to satisfy the following requirements of an environment. First, agents have to adapt to the information perceived, e.g. to resource limitations detected in the environment. In a similar way they may adapt to requirements *anticipated* from the perceived information. Adaptation can further take place in reference to goal-based constraints as well as in reference to a change of goals. Finally, in some cases, agents can adapt their strategy of interacting with other agents which might happen on demand or by opportunity. Examples of environments producing the mentioned requirements are given below with reference to the decision models introduced in Chap. 2.

- Recall the CLSP formulation of Example 2.6. If the available time Q is occasionally enlarged or reduced in a certain period, this information has to be perceived by agents which are responsible for scheduling the production program. In case that the period in question lies in the future the information still may be helpful in order to determine queuing decisions with respect to an anticipated relaxation or bottleneck.

- The requirement of adaptation to modified or changed goals is illustrated by the CVRP formulated in Example 2.7. Here the objective is to minimize the total distance of routes using K vehicles. A goal-based constraint modification refers to a changed number of vehicles available. Accordingly, a changed goal can express e,g, the desire to reduce the average delay of the vehicles reaching the costumers.

- The adaptation of strategies was first approached by Axelrod (1987) in a game-theoretic context[3]. From the organizational viewpoint, adaptation of interaction strategies requires an independent coordination of agents, compare Chap. 2(3.2). Consider a problem like the CVRP where entities (customers) may be exchanged between other entities (vehicles). Agents responsible for routing the vehicles can completely assess the saving of each particular costumer exchange offered by another agent (recall that in contrast additional costs of

[2]This definition should not be confused with the weaker concept of *partly intelligent agents*, saying that an agent is equipped with a problem-specific heuristic, compare Mertens et al. (1996).

[3]For an economic interpretation of the resulting agent behavior see Dawid (1996).

inventory-holding arise from such exchanges in the CLSP). Therefore routing agents might independently change their own bidding strategy within the contract net of an exchange market.

Throughout the remainder of this section we compare the properties (i)-(iv) with the properties of GAs. First we outline the basic principles that allow GAs to sense and to act in an environment. Afterwards we point out their communication interfaces. This is a central prerequisite for using GAs as internal methods of interacting agents. The possibility to use EAs for the design of learning agents, addressed by property (v), is considered later on.

1.2 SENSING AND ACTING

Evolutionary computation has predominantly received attention because of its great usefulness in static function optimization. The interest in this field gave rise to advanced optimizers and insight into the kind of functions for which EAs are well suited. Nevertheless, the application of EAs is not limited to static environments[4]. This opens EAs for usage in agent technology.

In order to describe the way GAs sense a dynamic environment we take a look at their system structure. As outlined in detail in Chap. 2(1.2), adaptive systems build up an internal model of the environment. Based on a loose coupling, the internal model is continuously adjusted towards the observed conditions.

Fig. 5.1 illustrates that the population maintained by a GA can be interpreted like an internal model (compare Fig. 2.2). This internal model contains several samples of a search space which serve as a database for generating samples of increasingly better quality over time. In this way a population represents a *memory* for the entire system. The population is attached from two directions.

- Selection and reproduction copy existing samples into an intermediate population in order assemble new samples which then might be accepted to replace their originators. Both, copying and accepting of samples happen in reference to the goal specified. Hence, the system control is established together by the mechanisms of selection and reproduction.

---

[4]For this reason De Jong (1991) claims that for instance GAs are not function optimizers despite their important role in optimization. In review of the initial research he reminds at its motivating context, namely the intervention of robust systems for adaptation to dynamic environments. From this viewpoint optimization appears just as a special case of adaptation which applies if the environment keeps unchanged.

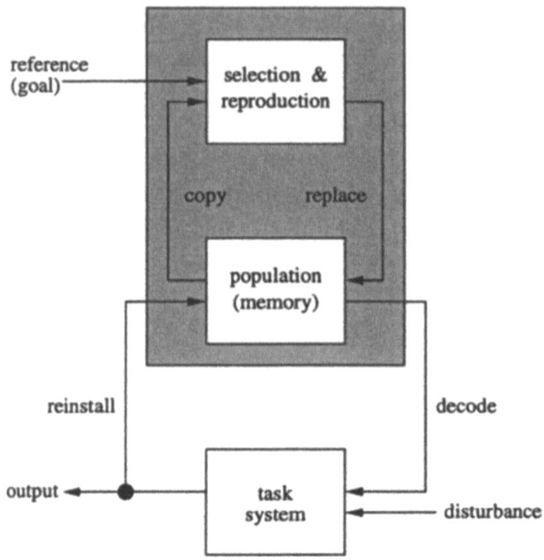

Figure 5.1 The external GA structure for sensing and acting.

- The task system is an object system, at which the sample population is tested regarding its overall quality of accomplishing the goal. This happens by decoding actions (*suggestions* how to perform the task specified) from the samples. Each action is simulated on a trial basis and the corresponding sample of the population is rewarded in terms of a fitness value.

If a task remains unmodified over time the described procedure is active until the involved GA has converged. The best performing action encoded in the final population then becomes implemented. In a changing environment the task system is influenced by disturbance which can confront the GA with modified conditions. As a result, samples existing in the population can loose compatibility with the current environmental situation (exhausted resources, failed anticipation, etc.) or they may require a re-evaluation in terms of fitness. Such effects are sensed while decoding a sample and testing the associated action in the task system. Necessary corrections in coding and reward of fitness are carried out during the simulation of these actions. Finally a compatible sample is reinstalled in the population. In this way the search space of a GA is regularly modified with respect to sensed information requirements.

The implementation of actions is driven by the task system. This procedure highly depends on the actual requirements observed. In a simple variant the best-valuated action which is compatible with the current conditions is chosen for implementation.

In the above described procedure of sensing and acting the GA can basically run within an open time horizon. Whenever an action has been implemented its effect in the environment is immediately perceived by the GA. This may result in a partial incompatibility of the currently existing internal model. Until the next action is requested, however, the GA still has time to reallocate new solutions for future demands. The main prerequisite of this ability is to keep a high degree of diversity in the GA population at any time. Presupposed that convergence control works well the GA has of course a good chance to fit the needs of the changed environment. In the meantime of two actions performed we therefore may call a GA an *anytime algorithm* which trades reasoning time against solution quality, compare Hayes-Roth (1995). In conclusion, GAs are principally capable of working in a permanent fashion.

By a detailed description of the way GAs sense and act in their environment we have shown that they have the ability to respond reactively to their environment. Since all actions provided have been valuated with respect to the goal pursued, we may call the overall system behavior goal-oriented as well. Moreover GAs provide control in the mode of a continually running process. To summarize, the potential actions of GAs are sufficient to meet the properties (i)-(iii) of agents.

1.3 COMMUNICATION

Communication of agents can serve one of the following two rationales. Agents may negotiate in order to decide on the distribution of common resources. Alternatively, agents may cooperate by transmitting individual results or information about future actions. The communication platform of negotiation models is usually realized on the basis of *blackboard architectures.* Cooperation basically depends on the principle of *message passing.* Since we concentrate on algorithms performing exploration tasks, negotiation models are not within our scope.

The technical question how messages can be passed to a GA was implicitly answered before. Communication interfaces of GAs are well defined by viewing them as ordinary adaptive systems. Through comparison of Figs. 2.3 and 5.1 two kinds of system interventions into the activity of GAs become obvious.

Goal Modification: the reference input of a GA is externally modified by means of a deformation of the fitness landscape it adapts to. In this way the goal pursued by the GA is effected.

Image Modification: the internal model of a GA is externally modified by means of structural changes of the samples it contains. In this way the memory of the GA is effected.

Example 5.1 The traditional hierarchical approach in production planning consists of first computing lot-sizes for a number of consecutive periods before sending this production plan to the scheduling level of the particular periods considered. The main disadvantage of this approach is that it cannot guarantee that a generated plan is feasible, i.e. that there exists a feasible schedule for all periods considered. We have seen in Example 2.6 that the capacity constraints taken into account at the lot-sizing level are rather naive as they merely state that machines cannot be occupied more than the time available in one period. A more reliable approach is to take the capacity absorption of a single item in the multistage production process into account[5]. However, unavoidable idle times of machines are still ignored in this approach because the actual sequencing of jobs on the machines is so far unknown.

A remedy promises a more detailed anticipation of the capacity demand. For this end the lot-sizing authority may apply an algorithm that consecutively simulates the processing of the lots assigned to the periods. This procedure typically starts with initial lot-sizes that correspond to the period demands of the products. If it turns out that this plan is feasible it certainly is optimal at the same time because inventory holding is completely avoided. Otherwise the initial plan exceeds the available capacity in at least one period. Then the scheduling algorithm reports the bottleneck situation to the lot-sizing authority by returning the backlog of work for that period.

If the algorithm was able to compute a feasible schedule, it returns the amount of free capacity in that period. Based on this information the lot-sizing authority has to revise the original production plan by either shifting splits of lots or whole lots from the bottleneck periods into earlier periods with unexploited capacities. Since each forward shift of items leads to an increase of inventory holding costs the lot-sizing authority should revise the current production plan with great care. Therefore a *goal-directed coordination* of the scheduling algorithm is required that slowly changes the initial plan. As a consequence diverse iterations of the multi-pass procedure are usually necessary in order to generate a feasible production plan of acceptable quality[6].

---

[5]This model extends the CLSP towards the multi-level capacitated lot-sizing problem (ML-CLSP). For a solution procedure derived from the heuristic of Dixon and Silver (1987) see Tempelmeier and Helber (1994).

[6]A methodology for the sketched multi-pass procedure has been proposed by Dauzère-Péres and Lassere (1994). In their approach the problem is solved by alternatively applying problem specific heuristics to the lot-sizing and the scheduling problems. In this way the boundaries of hierarchical planning are surmounted by an integrated planning approach. Unfortunately the approach is time consuming and allows to tackle relatively small problems only. The

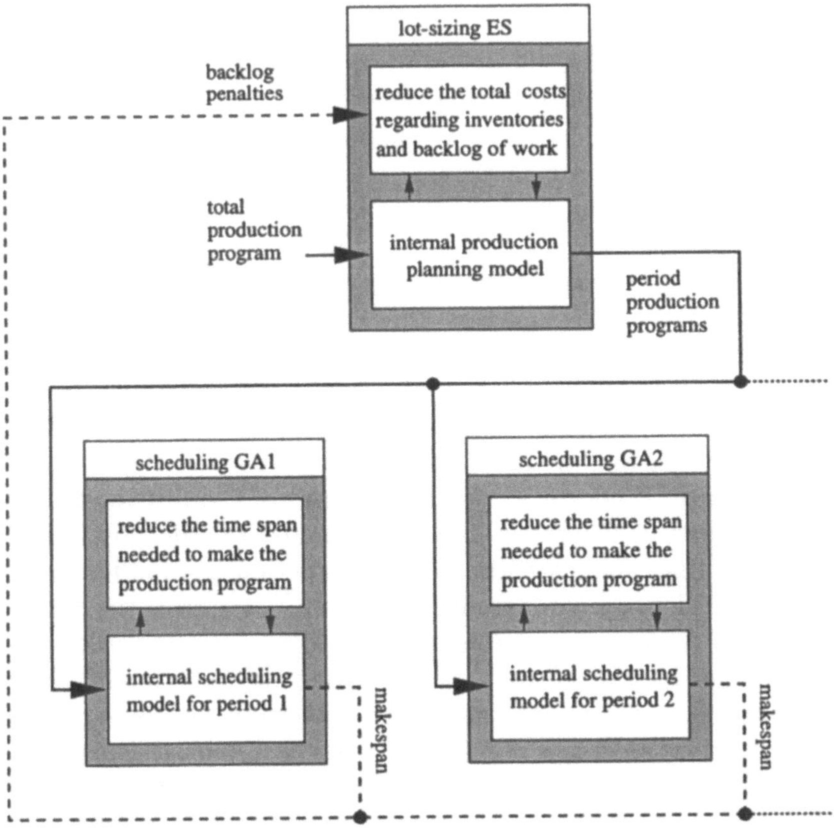

Figure 5.2. The communication structure of an evolution-based agent system for a two-level planning task (level 1 is lot-sizing and level 2 is scheduling). The bold arcs refer to image modifications of the scheduling GAs and the dashed arcs to a goal modification of a single-point ES used for lot-sizing.

Every machine scheduling algorithm can basically be applied in the above sketched organizational scheme of production planning. Different to problem specific heuristics, however, a GA is adjustable "on-the-fly" regarding changes initiated by the lot-sizing level. In order to point out the advantage of this approach let us assume that the period-related scheduling tasks are transferred to a number of independently operating GAs[7]. The involved GAs view the lot-sizing level as a common task system. To be more precise, each GA observes an excerpt of the task

---

computational results reported by Dauzère-Péres and Lassere (1994) refer e.g. to a lot-sizing problem consisting of six products and three periods.

[7]The design of GAs for scheduling is described in detail in Part II.

system for which it provides an internal model. These models consists of a number of solutions for the scheduling problem of each particular period. According to the above described multi-pass procedure the considered scheduling problems change over time by means of lots that are either inserted or excluded from the period production programs. A revision of the production plan is therefore passed to the scheduling level through an image modification of the concerned GAs. As a consequence the GAs need not necessarily to be started from the scratch. This feature promises an increase of the quality of planning as well as reduced computation times.

Thus far we have outlined the way an agent, say the lot-sizing authority, communicates with a GA in order to instruct this GA. Let us now concentrate on the way GAs respond to instructions received. Independent of the particular task performed, it is already clear that a GA can permanently generate an output in terms of the fitness of the best currently known solution. In the above model this output represents the minimal time span needed to make a period production program (*makespan*). Thus we can assume that every GA passes the makespan computed for its period to the lot-sizing authority after a preset number of generations are carried out. Alternatively the GAs might run until the adaptability of their internal model (measured by the convergence rate) falls below a certain threshold. After a GA has suspended running it waits for an intervention by image modification which causes the continuation of search again. In this way a detailed anticipation of the capacity demand can be initiated at the lot-sizing level for each of the periods. If we view the lot-sizing authority as an agent, we recognize that it actually coordinates the involved GAs in a goal-directed fashion.

Let us finally assume that an EA is responsible for the lot-sizing task by operating on the representation space of this problem[8]. Fig. 5.2 shows the overall communication structure of such a system. In the simplest case we can apply an Evolution Strategy with just one parent and one offspring (single-point ES). The parent solution represents the currently suggested production plan. The impact of this plan is investigated by several GAs which generate detailed production schedules for the periods under consideration. Dependant on the results produced by the scheduling GAs, a modification of the production plan is produced by the ES. Every new solution (derived from a mutation of the current solution) varies the existing plan slightly. Again, the modification is passed

[8]For a binary representation space of the CLSP see Helber (1994). This approach confines forward-shifts of product quantities to complete lots. An integer representation space which also permits lot-splitting is described by Fahl (1995).

to the scheduling level and the ES awaits a response for the periods in questions. Eventually, a new makespan is anticipated for these periods. This data enters the evaluation of the new plan in order to decide on a revision of the current plan provided by the ES. In other words, the assumed backlog of work in the periods causes a *goal modification* at the lot-sizing level, i.e. the fitness landscape of the lot-sizing problem is dynamically deformed in order to guide the search into feasible regions, compare Chap. 3(2.4). Unfortunately, it is not completely clear by now how suitable penalties of infeasible plans can be calculated as a function of the backlog observed[9]. ∎

Despite the technical description of message passing interfaces, the incorporation of EAs in an agent system requires detailed insight into the way the network is organized. By the above example we recognize that EAs are principally able to operate in such a network as demanded by property (iv). In reference to property (v), the next section shows how EAs can cope with changing environments.

2. ADAPTATION TO CHANGING ENVIRONMENTS

Adaptation to static environments is commonly equated with optimization. In difference, one could designate adaptation to changing environments as learning. This section starts with a brief review of previous applications of EAs to problems that change over time. Having described EAs as agents thus far, we then turn to a more careful distinction by concentrating on agents that incorporate EAs in order adapt automatically to an environment. From this viewpoint EAs stand proxy for a learning algorithm which is used by the agent. A rough draft of the prospects and limitations of machine learning is given before the function of evolution-based learning is analyzed.

2.1 PREVIOUS APPROACHES

From the viewpoint of decision models, a changing environment is either stated in terms of an objective function that varies over time or in terms of varying problem constraints or data. From the viewpoint of evolutionary search these kind of changes have different effects. Changes of the objective function result in a deformation of the fitness landscape whereas changes of constraints or data can additionally result in a reconfiguration of the underlying problem representation space. Both is

[9]As we will see soon, in the context of other problems goal modification of EAs is a well known principle for perceiving changes in the task system.

well distinguished in reference to the principles of goal modification and image modification.

Studies which investigate EAs in dynamic environments predominantly deal with goal modification. Typically a one-dimensional fitness function is considered for a certain number of generations before that function is modified more or less strongly. In order to accommodate such environmental changes Goldberg and Smith (1987) propose to expand the memory capacity of a GA. According to the biological model of *diploidy* they investigate a string representation being twice as long as necessary in order to encode the problem under consideration. The elimination of conflicts in this highly redundant representation is assigned to an operator which decides on the *dominance* of the contained information. If the optimum switches between two different states in the environment, the diploid representation is likely to produce better results than an ordinary *haploid* representation. Unfortunately, if the environment oscillates between more than two states, the advantage drawn from a diploid representation decreases.

Instead of an explicit enlargement of the memory capacity of the population, other researchers aimed on increasing the diversity in the population effectively. This is a crucial aspect because the tendency of EAs to converge too rapidly reduces their ability to maintain information about regions of the search space that may become more attractive as the environment changes.

For instance Grefenstette (1992) proposes to modify the population management of a GA as follows. After each generation carried out a percentage of the population is replaced by randomly generated individuals. The intended effect is to maintain a continuous level of exploration while preventing a distortion of ongoing search. For the particular problem investigated by Grefenstette a replacement rate of 30% gives the best performance.

Another suggestion, discussed by Cobb and Grefenstette (1993), is designated as *triggered hyper-mutations.* The idea of this approach is to monitor the best individuals in the population over time. If this measure declines, it is a sufficient indicator that the environment has changed. The triggered hyper-mutations then essentially restart a completely new search. An attractive feature of this approach is that it emulates the standard GA in a static environment. Unfortunately, the mentioned measure is not necessarily an indicator for a changed environment and therefore it may fail to trigger hyper-mutations in case of changes which do not influence the level of fitness of the so far best individuals.

In a similar approach of Vavak et al. (1996) the crossover and the mutation operator are suspended during periods where the average performance of the algorithm worsens. In this situation a *tracking operator* is applied which carries out a variable range of local search steps from the locations in the search space where the individuals reside. Afterwards the algorithm resumes its typical iteration using standard genetic operators for search.

A dynamic approach in Evolutionary Programming is reported by Angeline (1997). For a number of dynamic environments created, ordinary EP is compared against self-adaptive EP which automatically readjusts the mutation variances. It turned out that the self-adaptive method performs worse if the change of the objective function is relatively large. Presumably, adapting the mutation rate along with the population impedes an accurate tracking of optima. A fine tuning of search parameters typically retard the reaction of search regarding a strong environmental change. Therefore Angeline concludes that more rough adaptive search methods should generally be preferred in changing environments.

The approaches described so far reflect variable requirements of an environment in reference to changes of the objective of a problem. A different requirement of an environment results if the constraints of a problem change over time. Such an approach is reported by Mori et al. (1996). They take the *time-varying knapsack problem* (TVKP) as an example.

Example 5.2 In the knapsack problem one has a set of n items i, each with an associated utility $u_i > 0$ and a corresponding load $a_i > 0$. The objective is to fill a knapsack with a subset of the items such that their total utility is maximized with respect to the load constraint b of the knapsack.

In the TVKP it is assumed that the load capacity of the knapsack changes dynamically over a period of T time units. Candidate solutions of the problem are represented as binary strings of length n such that the i-th bit position indicates whether item i is filled in the knapsack at $t \in \{1, \ldots T\}$ ($x_i(t) = 1$) or not ($x_i(t) = 0$). The objective is to maximize the utility function

$$U = \sum_{i=1}^{n} x_i(t) u_i, \tag{5.1}$$

such that for all $t \in \{1, \ldots T\}$ the following conditions hold

$$\begin{aligned} &\textstyle\sum_{i=1}^{n} x_i(t)\, a_i \leq b(t), \\ &x_i(t) \in \{0, 1\}. \end{aligned} \tag{5.2}$$

Note that in this model the utility and the load of the items are kept constant while only the knapsack capacity $b(t)$ changes over time. Therefore the utility function is actually time independent.

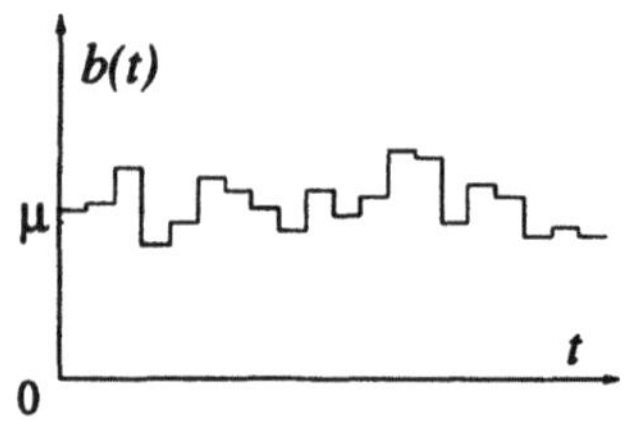

Figure 5.3 Time-varying knapsack capacity.

An irregular varying environment is achieved by modeling the capacity $b(t)$ as an autoregressive process with mean μ and variance σ^2 respectively. For $\mu = 0.5 \sum_{i=1}^{n} a_i$ and $\sigma^2 = 0.2\mu$ we set $b(0) = \mu$ and iteratively calculate $b(t) = 0.5 \cdot (\mu + b(t-1)) + 0.75 \cdot \sigma^2 \eta$ for $t > 0$, where η denotes a noise of zero mean and unity variance. The resulting change of available capacity is shown by example in Fig. 5.3.

In order to apply GAs to the TVKP Mori et al. (1996) introduce a fluctuation interval τ saying that a capacity change is perceived by the GA in every τ-th generation. According to the handling of constraints described in Chap. 3(2.4), the fitness of a solution is determined by its total utility and an additional term which penalties the fitness in case that the solution exceeds the knapsack capacity. This leads to a fitness function

$$f = U + v \cdot \min\left(0\,,\, b(t) - \textstyle\sum_{i=1}^{n} x_i(t) a_i\right), \tag{5.3}$$

where $v > 1$ denotes a violation weight. In order to avoid negative fitness values a further linear scaling of the fitness is applied.

It is reported that a standard GA produces feasible solutions of good quality for the TVKP if the fluctuation interval is set to a relatively short length. If the fluctuation interval exceeds a certain limit, however, the GA starts to converge which impedes further adaptation. ■

The example illustrates how GAs can be applied to an optimization problem where a capacity constraint is subject to changes that entry at discrete points in time. In order to restrain the GA from generating infeasible solutions the principle of goal modification is applied. A different way to manage varying conditions results if the principle of image modification is used for perceiving the environment. The solutions contained in the population are reallocated in a single step now. This idea becomes clear if we take a look at another variant of the TVKP.

Example 5.3 Let us now view the capacity of the knapsack as constant over time. In difference to the previous example we assume that the knapsack has to be filled with items taken from a true subset of the n items available. This situation may result from a time-varying demand of the items such that the set of the relevant items changes slightly in time. We can model this problem by assigning each item a probability $p_i(t)$ to be of relevance at time t. At each point in time t, the knapsack problem is

1. Determine the relevant items through probabilistic draws from the set of all items, and

2. Fill the knapsack with items taken from the set of relevant items such that the total utility is maximized.

For simplicity we can assume that all n items is assigned an identical time-independent probability p. If this uniform probability is set to $p = 0.9$ there is an average total of $0.9 \cdot n$ relevant items at each point in time.

Note that equations (5.1) and (5.2) still hold if the utility and the load of the excluded items are set to zero. For this reason a standard GA can handle the dynamic problem with respect to the principle of image modification. While decoding the binary strings, bit positions corresponding to items which are currently of no relevance are simply ignored. Alternatively, the string representation can also be adjusted after each change of demand. Moving from t to $t+1$ then requires the deletion of items which are no longer part of the subset and the insertion of newly drawn items. ∎

The above example reflects a discrete decision model which can be solved dynamically by rearranging the underlying problem representation structure instead of explicitly modifying the objective function. Notice that in this case actually the set of feasible solutions slightly moves within the search space while the overall fitness landscape remains constant over time.

To summarize, automated decision making in a changing environment can concern either the objective function or the constraints of the decision model or even both at the same time. EAs offer in all cases sufficient interfaces for accepting the challenge because they are able to track "moving targets". The crucial aspect for success is to retain a reliable degree of diversity in the population. Whenever the EA starts to converge the algorithm reduces its ability to adapt to future states of the environment.

Of course, an evolutionary algorithm can always be started from the scratch or be disturbed by using hyper-mutations, or the like. From the viewpoint of the agents reaction time, however, it is more efficient to carefully control the convergence of the learning algorithm and suspend running whenever the entropy of the population falls below a threshold.

2.2 MACHINE LEARNING

Learning is generally defined as the capability of a system to make structural changes to itself with the goal of improving its overall performance. Thereby learning is viewed as the result of an internal process which is based on feedback regarding the success or the failure of the system's behavior.

Learning processes of agents typically attempt to mimic human learning in terms of programs which process prior knowledge like facts, rules or constraints. The ability of an agent to learn corresponds to the ability to perform a given task increasingly more efficient over time. Still, human learning and machine learning remain distinguishable because human learning also includes the ability to perform new tasks which could not be performed before. This conjectured limitation of machine learning is explained by Michalski and Kodratoff (1990).

> One of the most striking difference between how people and computers work is that humans, while performing any kind of activity, usually simultaneously expand efforts to improve the way they perform it. This is to say that human performance of any task is inseparably intertwined with a learning process, while current computers are typically only executors of procedures supplied to them.

Several algorithms for machine learning are known from literature. Before we give a rough classification of the various techniques we take a look at the predominant theories on human learning. Psychology distinguishes three main types of learning, namely

- analytical learning (explanation-based learning),
- reinforcement learning (learning by conditioning), and
- inductive learning (learning by examples).

Independent of the particular kind of theory which has inspired a computational learning system, its architecture must allow to perceive the effect of prior activities. Providing suitable feedback mechanisms which enable a program to make useful changes to itself is therefore a primary issue of machine learning[10]. For this purpose knowledge representations have to be developed which are suitable regarding the implementation of structural changes.

Machine learning as suggested by traditional AI aims at making structural changes by extending the knowledge base of rule-based systems.

[10] As already remarked in Chap. 1, there has been an enormous growth of interest in this field over the last three decades. For a comprehensive overview and a classification of diverse machine learning approaches the reader is referred to Michalski and Kodratoff (1990).

Diverse methods derived from the principles of analytical learning can be employed. Yet most common is inference making according to axioms or to analogies.

Recent approaches predominantly employ CI-techniques for learning tasks. Probably most widely used are *connectionist learning* models. The computational framework of ANNs basically consists in the adjustment of connection weights between the processing units of a net, see Chap. 1(2.1). Since the observed error is back-propagated through the net, this concept shows conspicuous similarities with psychological principles of reinforcement learning.

Learning by evolution has a different flavor that rather follows the rationale of inductive learning, i.e. to derive general concepts on the basis of well known examples. The early draft of evolutionary learning prepared by Holland et al. (1986) is quite consistent with the design of Classifier Systems. As we have seen in Chap. 1(1.3), Classifier Systems extend the focus of traditional rule-based systems towards the generation of new rules. Recently a more universal understanding of evolutionary learning came up. E.g. Pesch (1994) treats learning within the framework of local search. Even more generally, Geyer-Schulz (1995) investigates learning of problem solving procedures over context-free languages.

2.3 LEARNING BY EVOLUTION

In order to assess the capabilities of agents exploiting the potentials of EAs, evolutionary learning is figured out in this section. We analyze the way EAs *learn* to perform a task by first disclosing their feedback mechanism and second clarifying the kind of structural changes they do.

FEEDBACK MECHANISM

Following De Jong (1990), insight into evolutionary learning is attained by focusing on the control of GAs. We have already sketched their external control structure in Fig. 5.1. Let us now turn to the internal control structure shown in Fig. 5.4. The interrelation of selection and reproduction describes the evolutionary dynamic in terms of a domain independent goal-seeking system. As outlined in Chap. 3(3.2), selection drives the passive reproduction process.

A system goal is reflected in this model by an externally given fitness landscape. This surface responds the quality of the agent's suggestion of how to perform a certain task. If we look at learning as tracking attractive areas in a landscape, we aware that the selection control will immediately react whenever the pursued goal or the constraints have changed in the environment. The common way of monitoring this process is to plot the average fitness of the internal model over time, or,

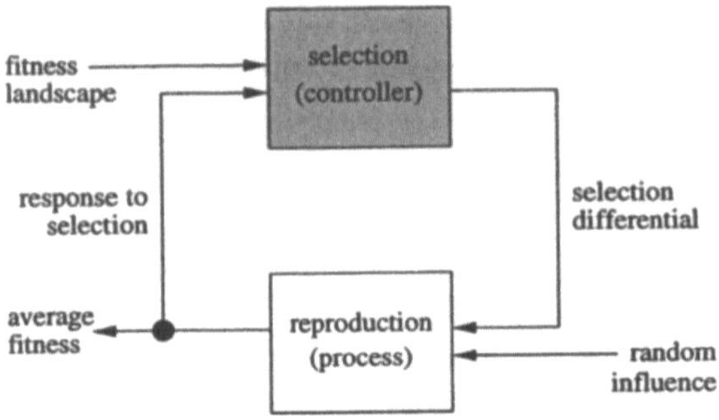

Figure 5.4 The structure of learning by evolution.

alternatively, the fitness of the currently best suggestion. Examples of such "learning-curves" are given in Fig. 4.3.

LEARNING OPERATORS

From the above viewpoint there is emphasis on evaluating learning in terms of changes in performance but not on capturing learning in terms of changes in memory. Let us now view reproduction as an active operation which produces innovative contributions. This aspect basically addresses the way a problem is represented. According to the classification of EAs, possible solutions to a problem may encode[11]

- a set of numerical parameters,
- static data structures (e.g. sequences or groupings), or
- variable data structures (e.g finite graphs or programs).

Different problem representations obviously require different ways of decoding a string into instructions for the task system. Basically, however, there is no difference between a task system which receives instructions by interpreting a real-valued vector or an agenda like a permutation, and a task system for which a computer program is executed. In order to adapt to the environment the EA takes changes of its memory in every case. We have described the common genetic operators earlier with respect to their characteristic of biasing a search process under static conditions. Under dynamic conditions the underlying fitness landscape slowly deforms. Thus, the suitability of bias can be validated by an assessment of its structure as well.

To summarize, we have seen that EAs are able to build up a memory, to reason on that memory and to adapt the memory in order to take

[11]The first item applies for ordinary GAs and ESs whereas the second item points to order-based GAs and to approaches in EP. Algorithms belonging to the last item address Genetic Programming and Classifier Systems.

self-improvements. Now everything is well prepared and we can depict the architecture of evolutionary adaptive agents by means of a proper framework.

3. THE ARCHITECTURE OF ADAPTIVE AGENTS

The notion of adaptive agents has received attention since Holland (1995) used it for analyzing complex systems. Without invoking any specific context, Holland determines agents by a collection of rules. This approach can be extended towards the consideration of other structures (e.g. numerical parameters) without loss of generality. The proposed framework consists of three components, an algorithm for *knowledge discovering*, an *evaluation system* called credit-assignment, and a *performance system.*

The architecture of adaptive agents described in the following puts Holland's framework in more concrete terms. We concentrate on the design of agents supporting the management of logistics operation. Due to the combinatorial nature of the addressed optimization problems, two further components are added in our model, namely a *problem representation system*, and an *encoding-decoding system.* A closer look at the architecture of adaptive agents will give us a good idea of what has been gained so far.

PERFORMANCE SYSTEM

The performance system specifies the agent's capabilities at a certain point in time, i.e. what the agent potentially could do without further adapting itself. It consists of a *detector* for perceiving information, a *knowledge base* containing a collection of suggestions how to instruct the task system and an *effector* for implementing actions.

The detector represents the agent's capability of perceiving different kinds of information. The knowledge base is viewed as the result of processing this information. Knowledge is given in form of *hypotheses* which express expectations concerning the impact of actions. The effector is responsible for selecting favorable hypotheses for implementation in the task system.

PROBLEM-REPRESENTATION SYSTEM

This system provides a decision model for the task system under consideration. It extracts a detailed parameterization for this model from information detected in the environment. Whenever the state of the

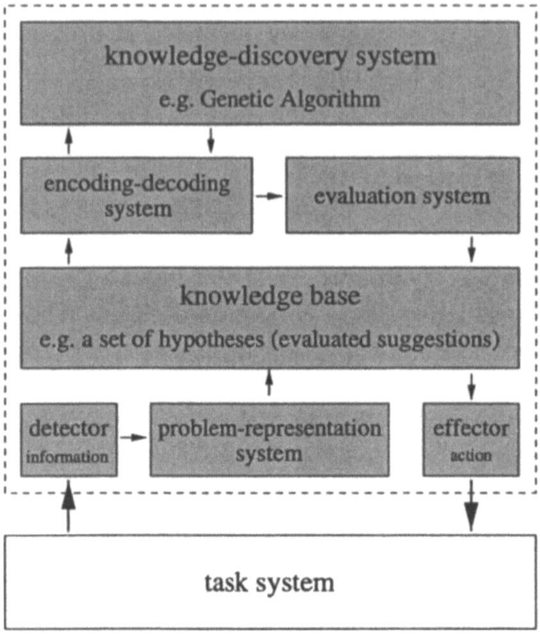

Figure 5.5 The architecture of evolutionary adaptive agents

task system changes, this immediately causes a parameter modification in the decision model.

Every parameterization of the decision model determines (i) the structure of inputs to the decision model and (ii) a function for evaluating the possible inputs by means of fitness. The problem representation system therefore specifies the agent's capability of perceiving information by image modification as well as by goal modification. If only parameters of the evaluation function have changed this information is passed on to the evaluation system. Otherwise the current hypotheses of the knowledge base are transferred into the new representation structure. In this way all suggestions remain feasible at any time.

EVALUATION SYSTEM

The rationale of assessing suggestions by fitness values is to provide the agent with an anticipation of the consequences of each particular action. This is realized by first simulating the outcome of a suggestion under the current state of the task system, and second by applying the evaluation function to this outcome. In this way reasonable hypotheses are generated. Since the quality of a suggestion depends on the current state of the task system, the existing hypotheses are re-evaluated automatically whenever parameters of the decision model have changed externally.

ENCODING-DECODING SYSTEM

Every suggestion contains a detailed instructions which can be immediately implemented in the task system (e.g. a schedule in a manufacturing system). The knowledge-discovery system yet depends on an aggregated representation (e.g. strings of fixed length over a finite alphabet) to realize syntactical manipulations. The encoding-decoding system takes care of this translation function. It supplies the knowledge-discovery system with an encoded versions of suggestions. The evaluation system is supplied in opposite direction by decoding strings into suggestions.

The decoding component of the system can additionally incorporate an iterative improvement heuristic or a problem-specific heuristic. This is to take improvements of a hypothesis which might be discovered while simulating the corresponding action in the task system. Improved hypotheses are reformulated in the knowledge base.

KNOWLEDGE-DISCOVERY SYSTEM

The generation of plausible hypotheses aims at the combination of building blocks taken from already tested suggestions. "Reasoning" of agents incorporates their past experiences in this way. The particular method used for recombining building blocks draws conspicuously on the design of genetic operators. The framework of Holland (1995) actually considers knowledge acquisition as a more pervasive process which is for instance also described by the learning algorithms developed in neural computation.

Notice that the above components do not require a strict synchronization in order to ensure the reliability of the agent. Just like the entire agent is concatenated with the task system, its components are viewed as being loosely coupled.

II

APPLICATIONS OF EVOLUTIONARY ADAPTIVE SYSTEMS

Chapter 6

PROBLEM REPRESENTATION IN LOGISTICS SYSTEMS

The idea of machine learning basically raises three questions. How can we represent knowledge concerning a particular task? How can we derive plausible hypotheses from the knowledge available? And finally, how can we experience any hypothesis in order to extend the knowledge base? As we have already seen, the latter two questions are answered by the paradigm of simulated evolution. However, for a specific problem under consideration the representational issue still matters. Representation is widely recognized as a central prerequisite of performance in evolutionary computation. This means that not any representation fits the need, but that a qualified problem representation has to be found, in order to generate promising hypotheses by a learning algorithm.

In this chapter we analyze representation structures for problems closely related to the management of logistics systems. The addressed issues have at their core extremely difficult to solve combinatorial optimization problems which, typically, can be traced back to a number of base problems. We take a short look at these base problems and point out their mutual relationships in first section. As it turns out that solutions of the base problems are representable within a common scheme, we propose a unified encoding system for adaptive agents afterwards. Further domain knowledge is included by the design of the decoding system and the knowledge-discovery system. The decoding system may incorporate local-search based algorithms as well as problem-specific heuristics whereas the knowledge-discovery system can employ different kinds of syntactical search mechanisms. The effectiveness of preserving encoded information by mutation and crossover mechanisms is investigated in a third section with respect to the various base problems.

1. COMBINATORIAL BASE PROBLEMS

Logistics management combines decisions of grouping (what), assignment (where) and sequencing or scheduling (when) of specific objects. Such complex tasks address e.g. cross-docking, the integration of inventory and transportation, container packing, the management of pickup-and-delivery systems and vehicle routing with time windows. For a comprehensive overview of these problems we refer to Bramel and Simchi-Levi (1997). Presented in this section is a collection of simple yet difficult to solve standard models[1], which are considered as essentials of integrate optimization problems of the "what-where-when" type. While we concentrate on appropriate representations of the base problems it turns out that a wide range of optimization in logistics is basically amenable to a largely uniform treatment by adaptive agents.

1.1 TRAVELING SALESMAN

The TSP was already considered by a discussion of its fitness landscape in Chap. 4(2.4). Given a set of n cities and distances d_{ij} between any two cities i and j, find a route of cities $\pi = (\pi_0, \pi_1, \ldots \pi_n)$ of minimal length such that $\pi_0 = \pi_n$ and every other city is visited exactly once. This is as to minimize the cost function

$$C(\pi) = \sum_{i=0}^{n} d_{\pi_i, \pi_{i+1}}. \qquad (6.1)$$

Solving non-trivial instances in the range of $n = 1000$ to optimality presently requires several hours of computation time, see Johnson and McGeoch (1997). Since the problem is easy to formulate yet difficult to solve, it has become a comparative test field for heuristics. Almost every type of meta-heuristic for local search has been applied to the TSP, see e.g. Kirkpatrick et al. (1983) for an approach using Simulated Annealing, Dueck and Scheuer (1990) for a Threshold Acceptance algorithm, and Gorges-Schleuter (1989); Freisleben and Merz (1996) for two progressing EA approaches.

1.2 QUADRATIC ASSIGNMENT

In the *Quadratic Assignment Problem* (QAP) n facilities have to be assigned to n locations. Given the distances d_{ij} between locations and the flow intensities f_{kl} between facilities, the problem is to find an assignment $\pi = (\pi_1, \pi_2, \ldots \pi_n)$ of the facilities π_i to the locations i which

[1] all of them are *NP*-hard, compare Garey and Johnson (1979) for all except the last problem which was proven by Du and Leung (1990).

minimizes the cost function

$$C(\pi) = \sum_{i=1}^{n}\sum_{j=1}^{n} d_{ij} f_{\pi_i,\pi_j}. \tag{6.2}$$

The QAP plays a central role in location science and has important applications e.g. in factory layout. Notice that the TSP is a special case of the QAP which applies for identical flow intensities. Different to the TSP, the QAP is almost intractable for being solved to optimality if $n > 20$. Therefore research has paid much attention to local-search based heuristics. We refer to Burkhard and Rendl (1984) for an approach in Simulated Annealing, to Nissen and Paul (1995) for a Threshold Acceptance algorithm, to Taillard (1995) for Tabu Search and to Fleurent and Ferland (1994) and Merz and Freisleben (1997) for different kind of EAs.

1.3 BIN PACKING

The *Bin Packing Problem* (BPP) is stated as follows. Consider a finite set of items $i \in \{1, \ldots n\}$, each associated with a specific weight w_i. A constant $Q \geq \max_i w_i$ is given denoting the capacity of a certain type of container which is referred to as a *bin*. The problem is to find a packing order $\pi = (\pi_1, \pi_2, \ldots \pi_n)$ which minimizes the number of bins needed to pack all items. One can think of the packing procedure as assigning items to bins in the order they appear in π. If it turns out that the i-th item π_i still fits into the currently packed bin, the available capacity of that bin is reduced by w_{π_i} and a corresponding binary variable is set to $x_{\pi_i} = 0$. Otherwise we set $x_{\pi_i} = 1$ in order to indicate that a new bin is started to be packed and its initial capacity is set to $Q - w_{\pi_i}$. Thus the BPP is as to minimize the cost function

$$C(\pi) = \sum_{i=1}^{n} x_{\pi_i}. \tag{6.3}$$

The BPP receives attention because it forms a point of origin for load-balancing and graph-partitioning problems. It has been extended further towards the consideration of two and three dimensions in order to access cutting-stock and container-storage problems. Numerous heuristics are available which is why we only mention two evolutionary approaches. An early paper of Smith (1985) deals with two-dimensional bin packing. In a recent work genetic local-search is investigated for the BPP, see Falkenauer (1995) and Falkenauer (1996).

1.4 VEHICLE ROUTING

We have dealt with this problem in Example 2.7 under the abbreviation CVRP. Assume that a depot has m vehicles available to serve n

customers. The capacity of the vehicles is limited to Q units while each customer i demands q_i units of this good. Let $i=0$ denote the depot and d_{ij} the distance between any two locations $i, j \in \{0, 1, \ldots n\}$. Find a plan $\pi = (\pi_{1,0}, \pi_{1,1}, \ldots \pi_{1,n_1+1}\pi_{2,0}, \ldots \pi_{2,n_2+1}, \ldots \pi_{m,n_m+1})$ which allocates n_k customers to vehicle k in the order in which this vehicle visits its customers[2] so as to minimize

$$C(\pi) = \sum_{k=1}^{m} \sum_{i=0}^{n_k} d_{\pi_{ki}, \pi_{ki+1}}. \tag{6.4}$$

It can be seen that the CVRP is actually a special case of the BPP which integrates m different TSPs. Promising local-search algorithms for this problem are Simulated Annealing and Tabu Search but GAs as well, see e.g. Osman (1993) and Rochat and Taillard (1995). The CVRP is of direct relevance for the fleet management of haulage companies. Therefore minimizing cost-based criteria is of interest in practice rather than minimizing the sum of the route lengths. A GA approach for this task is e.g. reported by Kopfer et al. (1994).

1.5 SEQUENCING WITH TIME WINDOWS

Finally we consider a problem of sequencing n jobs with given time windows. In this problem each job has to be processed by a single machine for a prescribed duration p_i, called its processing time. For each job a ready date r_i and a due date d_i are announced which represent a time window for processing that job. No job can be started earlier than its ready date and it is desired to complete all jobs before their due date because tardy jobs lead to proportional penalty costs. Actually this problem is an *open* TSP where the distance between the last and the first job is not taken into account. Hence the problem is to find a processing sequence $\pi = (\pi_1, \pi_2 \ldots \pi_n)$ of the jobs that minimizes the total costs incurring from job tardiness[3]

$$C(\pi) = \sum_{i=1}^{n} \max(c_{\pi_i} - d_{\pi_i}, 0). \tag{6.5}$$

Note that the completion times of the jobs have to be calculated iteratively for this formula. The earliest possible starting time of a job is determined by the larger value of its ready date and the completion time

[2] Note that $n = \sum_{k=1}^{m} n_k$. For all vehicles $k \in \{1, \ldots m\}$ the locations $\pi_{k,0}$ and π_{k,n_k+1} refer to the depot.

[3] Given the projected completion time c_i of job i, its tardiness is calculated by $T_i = \max(c_i - d_i, 0)$.

of its preceding job on the machine. Assumed a fictive initial job has been completed at $c_0 = 0$ the further job completion times are given by $c_{\pi_i} = \max(r_{\pi_i}, c_{\pi_{i-1}}) + p_{\pi_i}$.

Sequencing with time windows represents a special class of general machine scheduling problems. Even though scheduling of manufacturing processes is discussed in greater detail later on, it is worth to mention here that the model has also important applications in vehicle routing supposed that due dates of delivery are prescribed by the customers.

2. ENCODING-DECODING SYSTEMS

Possible solutions of the base problems have been characterized by ordered lists of objects that either refer to cities, facilities, items, customers or jobs. This observation gives rise to a large unification of encoding systems for adaptive agents. Despite of the fact that many combinatorial problems are appropriately encoded by lists of certain objects, it cannot be guaranteed that every instantiation of such a list necessarily represents a feasible solution. The information acquired by an EA may require a post processing which decodes the problem representation scheme before a feasible solution is added to the knowledge base. Some decoding methods are therefore sketched for the base problems introduced.

2.1 ENCODING SCHEME

An ordering of the elements of a finite set is called a *permutation*. Permutations are widely used for the representation issue of combinatorial optimization problems. For instance the possible routes of a TSP or the possible facility layouts of a QAP are represented in a straightforward manner by permutations of either cities or facilities. This approach can be generalized in order to capture a wider range of combinatorial optimization problems.

A unique formulation of many problems can be described as follows[4]. Assume that a set of m resources is available in order to satisfy the demands of n objects. Find a permutation π which consecutively assigns objects to resources such that an overall cost function $C(\pi)$ is minimized. This process is represented by the permutation scheme

$$\pi = (\underbrace{\pi_1, \dots \pi_{n_1}}_{\text{resource } 1}, \underbrace{\pi_{n_1+1}, \dots \pi_{n_2}}_{\text{resource } 2}, \dots \underbrace{\pi_{n_{m-1}+1}, \dots \pi_n}_{\text{resource } m}). \tag{6.6}$$

Both parts of the general problem, the assignment part and the sequencing part, can appear alone. Obviously, if just a single resource

[4]cf. Bierwirth et al. (1996).

is available ($m = 1$), the permutation π is unpartitioned, i.e. there is no assignment part involved in the problem. On the other hand, if the objects demand resources exclusively ($m = n$), the permutation is m-partitioned which means that there is no sequencing part. Inside the range of these extremes we meet a class of problems with $1 < m < n$. Several problem types are well distinguished.

- We can deal with partitioning problems where each of a number of objects has to be assigned to exactly one out of a number of resources.
- We may consider partitioning problems which consist of grouping several objects and assigning those groups to the resources. The number of partitions of π and their sizes result from the capacities of the resources.
- A similar case is observed if the actual capacity demand of a group of objects depends on their internal order.
- Finally there may be no grouping problem involved because all objects require dedicated resources. Objects yet have to be sequenced for the resources and these sequencing problems are possibly interrelated by dependencies existing between objects that use different resources.

Altogether the permutation scheme (6.6) provides the following classification.

i. Assigning of exclusive resources to several objects ($m = n$).

ii. Grouping of objects for a number of resources (variable $m \geq 2$).

iii. Grouping of objects with integrated sequencing (variable $m \geq 1$).

iv. Simultaneous sequencing for a number of resources (fixed $m \geq 2$).

v. Sequencing of objects for one resource ($m = 1$).

Examples of these cases are given by the base problems considered before. Most obviously, the QAP and the TSP belong to class (i) and (v) respectively.

The BPP is addressed by class (ii). Since the objective is to minimize the number of allocated resources, m is a variable parameter in this problem. This class further contains problems where the number of permutations is apparently fixed. Recall e.g. the knapsack problem considered in Example 5.2. This problem is certainly a grouping problem with $m = 2$. However, the parameter m seems to be fixed just because there is an *unlimited* resource that absorbs the items not filled in the knapsack.

The CVRP is captured by class (iii). Although a maximal number of resources (vehicles) is prescribed for this problem, it can turn out that less resources are actually needed. Since unallocated resources do not appear in the permutation we may view the parameter m as a variable.

Finally we classify the machine sequencing problem with time windows. This problem belongs to the same class as the TSP as it merely requires to sequence objects for one resource. Looking at the problem in a more general way we consider production as a multistage process which usually involves multiple machines of different functionality in order to process one job. If we specify a job as a complex which consists of a set of operations to be carried out by different machines, the overall problem is to sequence this set of objects simultaneously for the machines. Such type of problems are addressed by case (iv) of the classification scheme.

An analytical approach to assess the quality of permutation representations has been developed by Radcliffe (1991) under the name *Formae Analysis.* Actually Formae Analysis is a generalization of the Schema Theory of Holland (1975, 1992) which extends the concept of a binary encoding towards the consideration of higher-valued and order-based encodings. Like Schema Theory, Formae Analysis suffers from disintegration of the specific properties of fitness functions. Therefore it does not approve conclusions about the suitability of genetic operators fitting a representation without taking a specific problem into consideration. As a remedy Radcliffe and Surry (1995) propose to measure the *fitness variance* of formae (the observed fitness variance of solutions processing a certain forma) empirically. If a forma has a high fitness variance this will indicate that a selection mechanism cannot estimate the average fitness of the forma accurately. This conjecture is confirmed by comparative studies of different representations for the TSP.

2.2 DECODING PROCEDURES

Combinatorial problems represented by permutation encodings are amenable to the treatment of order-based GAs. This special class of EAs has been designed for problems where a solution is represented in a natural way by reading a permutation either as a sequence ($m=1$) or as a string of positions ($m=n$). Originally, less attention was paid to constrained permutation problems where a solution does not evolve until the permutation is decoded. Nevertheless it turned out that even such problems can be tackled successfully by EAs because the need for decoding provides a suitable interface for the incorporation of problem-specific heuristics. In this way adaptation enables the propagation of combinatorial domain knowledge which led to an increasing interest in *hybrid* EAs. For the base problems considered we review the most widespread decoding procedures in this section.

TSP AND QAP

We have characterized both problems as borderline cases of our encoding scheme because their possible solutions are represented directly as permutations. Hence the detour of indirect decoding is unnecessary in order to attain feasible solutions from the knowledge discovery system. It yet showed that adaptation alone is often too weak to produce competitive solutions. Local search can enhance adaptation by operating on the permutation strings after they have been generated by the discovery system.

Fortunately, the TSP and the QAP allow the definition of attractive neighborhoods. For the TSP we can estimate the outcome of a perturbation of a route within the permutation scheme in constant time. A slight perturbation refers to two positions i and j in the permutation π where $1 \leq i, i+1 < j < n$. The corresponding sub-route $(\pi_{i+1}, \pi_{i+2}, \ldots \pi_j)$ is removed from π and again inserted in inverse direction. For $j = i+2$ this neighborhood corresponds to the basic TSP neighborhood defined in Chap. 4(2.4). Assumed the problem is symmetric (i.e. $d_{kl} = d_{lk}$ for all cities k, l), two edges of the route are exchanged by one neighborhood move with associated costs of

$$\Delta C(\pi, i, j) = d_{\pi_i, \pi_{j+1}} + d_{\pi_j, \pi_{i+1}} - d_{\pi_i, \pi_{i+1}} - d_{\pi_j, \pi_{j+1}}. \qquad (6.7)$$

This formula represents the basis for the *n-opt* algorithms of Lin (1965). Simple iterative improvement algorithms for the QAP work similarly. Instead of exchanging edges in a route we now depend on exchanging facilities in their assignment to locations. The basic neighborhood of the QAP therefore refers to a pairwise exchange of facilities at positions i and j in the permutation π where $1 \leq i, j \leq n$. For symmetric distances between locations the costs associated with such a neighborhood move are calculated in linear time

$$\Delta C(\pi, i, j) = 2 \sum_{\substack{k=1 \\ k \neq i,j}}^{n} (d_{jk} - d_{ik})(f_{\pi_i, \pi_k} - f_{\pi_j, \pi_k}). \qquad (6.8)$$

If, however, the triangle inequality does not hold for a particular problem, equations (6.7) and (6.8) do not hold either. In this case more expensive calculations of ΔC are necessary, see e.g. Taillard (1995).

For other problems the benefit of local-search based decoding may not be worth the higher computational effort spent. As an example we mention the *permutation flow-shop problem* (PFSP) where n jobs are processed in *line-production* involving a fixed number of machines. The material-flow system connecting the machines does not allow that jobs overtake any other jobs. Thus, once a processing order of the jobs is chosen for the first machine, this order cannot be changed anymore. The optimization task formulated in the PFSP consists of finding a processing order π of the n jobs which minimizes the total makespan ($C_{\max}$). Like single machine sequencing problems this problem can be stated as an open TSP, which means that $m = 1$ holds. Unfortunately, the PFSP lacks an efficient neighborhood. For instance a pairwise exchange of jobs which transforms π into π' requires to build the solution corresponding to π' from scratch. Consequently the effect of a perturbation has to be calculated costly by $\Delta C(\pi, i, j) = C_{\max}(\pi) - C_{\max}(\pi')$, see e.g. Osman and Potts (1989). A non-hybrid GA therefore often produces better solutions of difficult PFSP instances in less computation time than an iterative improvement algorithm can, compare Bierwirth (1993).

BIN PACKING

For permutation problems neither captured by $m=1$ nor by $m=n$ the encoding scheme still offers a way of representing solutions indirectly. This is based on the observation that different solutions to such problems can be constructed, depending on the order in which the objects are handled. For this end a problem-specific procedure is employed which decodes a permutation of objects into a particular solution.

This idea is probably best illustrated for the BPP. Every instance of this problem can be solved by taking items one by one and pack them according to the rule *first-comes first-serve* (FCFS) into the current bin as long as the bin capacity is not exceeded. A way to improve this procedure is to replace FCFS by one of the well known heuristics *First-Fit* (FF) or *Best-Fit* (BF). FF starts with packing the first item in the first bin. Each further item j is packed in the lowest indexed bin whose current content does not exceed $Q-w_j$. BF works similar to FF except that it packs item j in the bin whose current content is the largest but does not exceed $Q-w_j$.

Further advantage can be taken from the rationale of the BF heuristic as proposed by Reeves (1994). Each time an acceptably full bin is produced by BF, this bin is fixed, i.e. the contained items are eliminated from the problem[5]. In this way promising bins are propagated which increasingly reduces the size of the search space. The computational

[5]Notice that if a GA applies this decoding procedure it effects an image modification of the population by its own.

results of Reeves show that the performance of the heuristic decoding procedure strongly benefits from the reduction approach.

VEHICLE ROUTING

Decoding procedures for the BPP basically assign objects to resources according to the order they appear in the permutation list without using any knowledge of subsequent objects in the list. This principle can be transferred directly to other grouping problems. For the CVRP a *heuristic decoding* has been approached by Kopfer et al. (1994).

As a simple variant a permutation decoding is tested where customers are assigned to vehicles via FCFS by taking care that the capacity constraints are not violated at the same time. Since this approach produces poor results a sophisticated decoding, which incorporates ideas of the famous *Savings Algorithm* due to Clarke and Wright (1964), is developed. In this approach all customers are initially projected to be served by exclusive vehicles that results in a corresponding number of pendulum-tours starting at the central depot. In the first step of the decoding procedure the possible saving-values for any two customers in the problem are calculated. Such a value expresses the saving expected from serving two customers consecutively by the same vehicle instead of serving them separately. Next, for each customer location, saving lists are built up containing the possible savings in decreasing order. Finally a solution is decoded from a permutation of customers in the following way. For the first customer in the list it is checked whether its highest saving can be realized without constraint violation. If this is possible the corresponding customers are grouped together for being served by the same vehicle. Otherwise the next-highest saving is checked, and so on. If either a saving has been realized or if no saving can be realized for a customer location, the decoding procedure continues to assign the next customer of the permutation list to a vehicle. The computational results reported by Kopfer et al. (1994) indicate that solutions within 1% of the optimum can be found for CVRP instances consisting of about 100 customers.

SEQUENCING WITH TIME WINDOWS

Decoding a permutation for a sequencing problem with time windows means to determine a temporal order in which a number of objects is assigned to certain resources. More precisely, starting times of allocating resources (say machines) must be scheduled for objects (say jobs) without overlapping in time. In the simplest case we read a permutation as a job sequence which directly represents the processing order for a single machine. This is done to determine starting times of jobs in

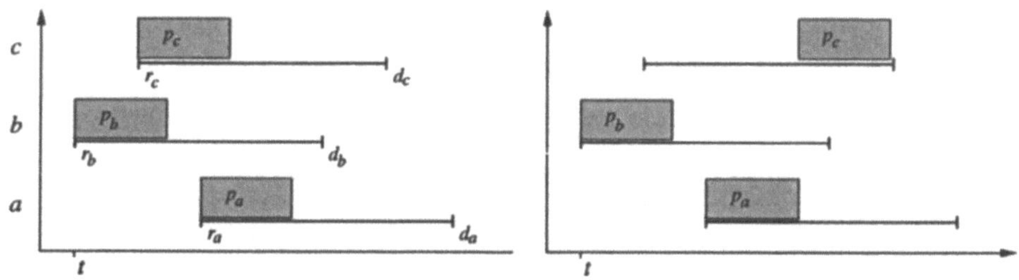

Figure 6.1. On the left three jobs a, b and c are given with a ready date, a due date and a processing time. While decoding permutation (a, b, c) on an active basis, job a gives priority to job b but not to job c as it is shown to the right.

FCFS manner at either their ready date or the completion time of the job scheduled before.

Like before the decoding procedure is easily strengthens if a problem-specific heuristic is incorporated. For illustration purpose we assume three jobs a, b and c as shown in the left chart of Fig. 6.1. Consider a permutation $\pi = (a, b, c)$. Obviously, determining a starting time for job a earlier than for jobs b and c (because both succeed job a in the permutation) means a waste of potential machine utilization which comes from the late ready date of job a. A remedy is promised by the well known scheduling algorithm of Giffler and Thompson (1963) which produces a so called *active* machine schedule. A decoding procedure adopted from this algorithm is:

1. Set time $t = 0$.
2. Compute the earliest possible starting time of the first job entry in π by $t_{\pi_1} = \max(t, r_{\pi_1})$.
3. Determine the leftmost job π_k in π which could be completed before t_{π_1}. This means that $t_{\pi_k} + p_{\pi_k} \leq t_{\pi_1}$ holds with $t_{\pi_k} = \max(t, r_{\pi_k})$.
4. If no such k exists schedule π_1 at t_{π_1}, delete π_1 in π, and set $t = t_{\pi_1} + p_{\pi_1}$. Otherwise schedule π_k at t_{π_k}, delete π_k in π, and set $t = t_{\pi_k} + p_{\pi_k}$.
5. If π still contains jobs goto Step 2. Otherwise terminate.

The above algorithm iteratively reduces the permutation π while decoding. At each point in time t the job occuring at first position in π is given the highest priority to be scheduled next. This priority is ignored if there exist so far unscheduled jobs which could be processed prior to the first job without delaying its projected starting time. Among those jobs the one at leftmost position in the permutation is selected for scheduling. Note from the right chart of Fig. 6.1 that the decoding algorithm in the example actually produces the job sequence (b, a, c) instead of (a, b, c).

2.3 REDUNDANCY AND FORCING

Looking at different decoding procedures it turns out that quite often various permutations actually encode identical solutions for a particular problem under consideration. Let us take a look at a small instance of the BPP consisting of three items a, b and c with weights $w_a = 0.5$, $w_b = 0.7$ and, $w_c = 0.4$ and a bin capacity of $Q = 1.0$. Consider permutation $\pi = (a, b, c)$ now. Decoding according to FCFS requires three bins because the bin allocated for item a cannot include item b, and the bin allocated for b cannot include item c. Consequently a third bin is allocated for item c. However, using a heuristic decoding like FF or BF always leads to a solution that merely needs two bins as items a and c can be packed together. In other words, in this small instance the possible six permutations all encode the same optimal solution if FF-decoding is used. Hence we see that the stronger the decoding heuristic is the larger the *degree of redundancy* of the encoding gets.

From the point of view of adaptive search a redundant encoding is generally conjectured as unfavorable because it enlarges the size of the space that has to be searched. Furthermore, if the permutation order is frequently ignored by the decoding procedure, the encoded information diffuses in the permutation scheme. As an example we look at the above permutation π again. Although items a and c are not placed together, FF-decoding groups both in the same bin. Radcliffe (1991) has shown that this phenomenon impedes a syntactical search operator to identify building blocks of the encoded solutions. If redundancy of solutions can be identified via an *equivalence relation* a search operator should effectively be able to "fold out" the redundancy.

Equivalence of permutations can be claimed with respect to the agenda defined in the permutation scheme (6.6). Recall that this agenda intends to write a permutation with a sequence of objects as they allocate resources. In other words, for every solution x of a problem which has been heuristically decoded from a permutation π_1 there exists an equivalent permutation π_2 from which x can be decoded in FCFS manner. According to this principle we may rewrite the above permutation π as (a, c, b) because it produces the same solution when FCFS-decoding is applied.

The idea to rewrite permutations with their resource allocation sequence has been introduced by Nakano and Yamada (1991) in the context of scheduling problems under the name *forcing*. Forcing substantially alleviates the problem of redundancy as it intensifies search by concentrating on a deputy subset of the entire permutation space. In this way forcing has two positive impacts: it speeds up the convergence rate of adaptation and it enhances the reliability of search at the same time.

3. LEARNING OPERATORS

The fact that there is a standard way of representing quite different combinatorial problems does not imply that these problems can be solved effectively by using "standard operators" for knowledge discovery. In fact it is well known that the genetic operators have to respect the ordering properties of a particular problem in order to evolve reliable hypotheses at high rates. In this section we first investigate the prospects and limits of representing domain knowledge by permutations. Afterwards we demonstrate the different capabilities of mutation and crossover techniques to preserve such knowledge.

3.1 ORDERING CATEGORIES

The success of an evolution-based learning process crucially depends on the interaction between the genetic operators and the underlying problem encoding. A particular operator works well for an encoding whose informational content it is able to maintain. This means that detailed characteristics of proven hypotheses are passed on with high probability. On the other hand an operator will fail for an encoding whose informational content it does not respect. In this case the characteristics of successful hypotheses cannot function as building blocks and the encoding is likely to mislead the knowledge-discovery system.

Using permutations for the transport of information can stress different properties such as the *absolute ordering* or *relative ordering* of the permutation elements. These categories produce a reference to the borderline cases of the encoding scheme presented in Sect. 2.1.

- An element of a permutation can express meaningful information with regard to the position it appears at. Pronouncing the *absolute ordering* is useful in assignment problems ($m = n$).

- Two elements of a permutation can express meaningful information if they are adjacent. Pronouncing the *relative ordering* is useful in sequencing problems ($m = 1$).

Obviously, for the TSP the relative ordering of permutations is crucial whereas for the QAP the absolute ordering is of importance. A more pervasive description of ordering which captures permutation problems in general is based on the concept of precedence. This is the fact that every element of a permutation has a precedence relation to every other element.

- An element of a permutation can express meaningful information with regard to the other elements (or a subset thereof) by either preceding or succeeding them in the permutation. Pronouncing the *global precedence* is particularly useful for combinatorial problems with $1 < m < n$.

Notice that every permutation defines its own Boolean precedence matrix. Consider e.g. element d in the permutation $\pi = (b, c, d, a)$. This element produces a binary vector $\binom{abcd}{100-}$ saying that d precedes element a but not b and c in π. The complete precedence matrix indicates the absolute ordering and the relative ordering of a permutation as well. Since there is just a single bit set in the above vector (i.e. d precedes only one other element), element d must occur at penultimate position in π. More general, if i bits are set in the precedence vector of an element, it occurs at position $n-i$ in the permutation, where n denotes the permutation length. From this it is clear that the binary vector associated with that element appearing directly to the left of d contains exactly one bit more than the vector associated with d.

For constrained problems the precedence matrix is viewed as a decision table for solving the conflicts which might arise concerning the allocation of resources. Consider the above permutation π again. It gives for instance preference to b against d whenever these elements take part in a conflict. Of course, whether or not such a conflict actually arises depends on the particular problem instance under consideration and on the decoding procedure used[6]. In some constrained problems there can exist (i) natural or (ii) no precedence relationships at all among some of the objects involved. In the CVRP a natural precedence between customers results for instance from the specification of time windows. In contrast, in a scheduling problem there often exist no precedence relation between two operations because they are allocating different resources. Here the global precedence defined by a permutation is partly ignored while decoding it.

Thus far we have described ordering categories of permutations capturing the properties of *absolute ordering*, *relative ordering* and *global precedence*. Recently the same classification has been proposed with regard to Formae Analysis. In the context of the PFSP Cotta and Troya (1998) distinguish between position formae, adjacency formae and

[6]Note that the resource allocation order of objects which are not involved in common conflicts depends on the decoding procedure alone.

precedence formae. By comparing the fitness variances of these formae they empirically verify absolute ordering being of predominant relevance. This confirms a previous result of Stöppler and Bierwirth (1992) made for the PFSP.

The ordering categories of permutations have different relevance for the combinatorial base problems considered before. We nevertheless expect from recombination to reliably process the informational content stressed in each particular problem class. In order to assess the bias of different genetic operators three measures are introduced. The content of information shared by two permutations π and π' of length n concerning

absolute ordering is determined by the number of positions $i \in \{1, \ldots n\}$ where $\pi_i = \pi'_i$. The ***normalized distance in absolute ordering*** d_a is this number divided by the largest possible value n such that $0 \leq d_a(\pi, \pi') \leq 1$ holds for arbitrary permutations,

relative ordering is determined by the number of pairs $i, k \in \{1, \ldots n{-}1\}$ where either $(\pi_i, \pi_{i+1}) = (\pi'_k, \pi'_{k+1})$ or $(\pi_i, \pi_{i+1}) = (\pi'_{k+1}, \pi'_k)$[7]. The ***normalized distance in relative ordering*** d_r divides this number by the largest possible value $n-1$ such that $0 \leq d_r(\pi, \pi') \leq 1$ holds,

global precedence is determined by the Hamming Distance of their precedence matrices. The ***normalized distance in global precedence*** d_p divides this number by the largest possible Hamming Distance $(n^2 - n)/2$ such that $0 \leq d_p(\pi, \pi') \leq 1$ holds.

Example 6.1 The above measures are illustrated for $n = 9$ using

$$\begin{aligned} \pi &= (a, b, c, d, e, f, g, h, i), \\ \pi' &= (b, c, i, d, e, h, g, f, a). \end{aligned}$$

1. Since both permutations have identical elements at positions $i = 4, 5,$ and 7 their normalized distance of absolute order is $d_a = 3/9 = 0.\overline{3}$.

2. Both permutations have identical adjacent elements (b, c), (d, e) at positions $i = 2, 4$, $k = 1, 4$ respectively. Furthermore, at positions $i = 6, 7$, $k = 7, 6$, there are identical inverse-adjacent elements (h, g), (g, f). Hence the normalized distance of relative order is $d_r = 4/(9 - 1) = 0.5$.

3. The associated precedence matrices of π and π' are given below.

[7] To meet conditions of asymmetry the consideration is restricted to the first term.

π	a	b	c	d	e	f	g	h	i
a	–	1	1	1	1	1	1	1	1
b	0	–	1	1	1	1	1	1	1
c	0	0	–	1	1	1	1	1	1
d	0	0	0	–	1	1	1	1	1
e	0	0	0	0	–	1	1	1	1
f	0	0	0	0	0	–	1	1	1
g	0	0	0	0	0	0	–	1	1
h	0	0	0	0	0	0	0	–	1
i	0	0	0	0	0	0	0	0	–

π'	a	b	c	d	e	f	g	h	i
a	–	0	0	0	0	0	0	0	0
b	1	–	1	1	1	1	1	1	1
c	1	0	–	1	1	1	1	1	1
d	1	0	0	–	1	1	1	1	0
e	1	0	0	0	–	1	1	1	0
f	1	0	0	0	0	–	0	0	0
g	1	0	0	0	0	1	–	0	0
h	1	0	0	0	0	1	1	–	0
i	1	0	0	1	1	1	1	1	–

Precedence matrices are always inverse-symmetric. Consequently there is a total of $(9^2 - 9)/2 = 36$ non-redundant entries to compare in our example. The Hamming distance of the matrices in the upper (or lower) triangle yields a value of 16. Hence the normalized distance of global precedence between π and π' is $d_p = 16/36 = 0.\overline{4}$.

■

The example verifies that a correct interpretation of the difference in two permutations depends on the ordering category stressed by the problem under consideration.

3.2 MUTATION TECHNIQUES

The central idea of mutations is to introduce substantially new information into the knowledge base by chance. Unfortunately, mutations often fail to discover useful information. Therefore it appears highly desirable to provide mutation operators that perturb an existing hypothesis in the slightest possible way. By carefully increasing the mutation rate, an operator with this quality enables a fine-tuned perturbation of a hypothesis even if singular mutations are too weak to effectively improve the performance of the adaptive agent[8]. Designing mutation mechanisms with the mentioned quality, however, requires insight into the change of informational content of a hypothesis as it is effected by a certain syntactical perturbation.

For permutation encodings various mutation techniques are proposed in literature. In our investigations we follow Syswerda (1991) who considers three basic techniques which he refers to as *scramble mutation*, *order-based mutation* and *position-based mutation*. Unfortunately, the

[8]cf. the recombination control model in Chap. 4(1.3).

Table 6.1. Average normalized distances between permutations and mutants.

mutation technique	d_a	d_r	d_p
sublist inversion	0.31	0.02	0.148
random reinsertion	0.34	0.03	0.007
pairwise exchange	0.02	0.04	0.013

naming of these techniques seems to suggest the way they respect different ordering categories. Therefore we rename them which shall rather indicate the syntactical operation a mutation technique is based on.

In order to mutate a permutation of length n, we first select two positions $1 \leq i < j \leq n$ randomly. In scramble mutation the permutation sublist $(\pi_{i+1}, \pi_{i+2} \ldots \pi_j)$ is next extracted from π, then scrambled, and finally reinserted again. We concentrate on a variant of this method, called *sublist inversion*, which actually reinserts a permutation sublist in inverse direction instead of scrambling it before. In *random reinsertion* merely the element π_j is extracted from π and placed before π_i (position-based mutation). A *pairwise exchange* of elements simply interchanges the positions of π_i and π_j in the original permutation (order-based mutation). Examples of the techniques are given in Fig. 6.2 for $n=9$, $i=2$ and $j=7$.

	$\pi\ = (a, \underline{b}, c, d, e, f, \underline{g}, h, i)$
sublist inversion	$\pi' = (a, b, \underline{g, f, e, d, c}, h, i)$
random reinsertion	$\pi' = (a, \underline{g, b}, c, d, e, f, h, i)$
pairwise exchange	$\pi' = (a, \underline{g}, c, d, e, f, \underline{b}, h, i)$

Figure 6.2 Mutation operators.

The described mutation techniques basically correspond to elementary neighborhoods which can be defined for any permutation problem. To investigate how well the techniques respect particular ordering categories we apply the mutation operators to 1,000 random permutations of length $n = 100$ each. The normalized distances in absolute and relative ordering and in global precedence between original permutations and their mutants are computed. This measure, averaged over the 1,000 samples, is reported in Tab. 6.1.

As one could have expected the operators respect the ordering categories to a fairly different extent. Sublist inversion and random reinsertion bias the absolute ordering at about one third whereas pairwise exchange modifies it just slightly. Note that the absolute ordering of permutations is even more touched by random reinsertion than by sublist inversion. In comparison, the impact on relative ordering is quite similar for all operators. Here sublist inversion appears little favorable at least if symmetry holds for adjacent elements. Global precedence information of permutations is best preserved by random reinsertion because at most the precedence relations of a single element will be swapped. In pairwise exchange the precedence relations of two elements are touched and in sublist inversion even more elements are involved. Consequently these operators are less respectful regarding global precedence.

To summarize, every mutation technique dominates one ordering categories and performs worst for another. This gives evidence to the frequently made observation that one operator works significantly better than others in a certain problem class (e.g. sublist inversion for symmetric TSPs). Nevertheless, a definite decision for a mutation technique should finally depend on some computational experience. If the problem encoding is highly redundant a majority of mutants actually will produce identical hypotheses. In such a situation the effects of an operators might be alleviated by the decoding procedure used.

3.3 CROSSOVER TECHNIQUES

Successful knowledge discovery of an evolutionary search process depends on a balance of exploration and exploitation. From this viewpoint the following design guideline for crossover operators is taken into consideration.

i. The crossover technique must respect the representation scheme.

ii. It should be able to maintain encoded information reliably.

iii. Information of hypotheses is passed on in similar proportions.

iv. To avoid an overlay of mutations, the generation of information not contained in the employed hypotheses is evaded as far as possible.

Many crossover techniques for the permutation scheme are proposed in literature, compare e.g. Oliver et al. (1987) and Fox and McMahon (1991). Investigations on the suitability of an operator regarding demands (ii)-(iv) usually refer to a particular kind of optimization problem. Numerous studies compare for instance the performance of crossover operators in sequencing and scheduling problems, see Syswerda (1991); Kargupta et al. (1992) and Cotta and Troya (1998). In the context of this chapter, however, we ask more generally to what extent crossover can pay attention to the previously introduced ordering categories of permutations. For this end we concentrate on three crossover techniques which are designed to emphasize absolute ordering, relative ordering and global precedence differently.

The most common ways to perform a crossover of two permutations apply either *partially-mapped crossover* (PMX) or *order-based crossover* (OX). These operators are already described by Goldberg (1989).

parent 1	$\pi_1 = (a, b, \underline{c, d, e, f}, g, h, i)$
parent 2	$\pi_2 = (f, b, e, a, i, c, g, d, h)$
partially-mapped	$\pi_o = (\ , b,\ ,\ ,\ ,\ , g,\ , h)$
	$= (i, b, \underline{\qquad\qquad}, g, a, h)$
	$= (i, b, c, d, e, f, g, a, h)$
order-based	$\pi_o = (\ , b,\ , a, i,\ , g,\ , h)$
	$= (b, a, i, \underline{\qquad\qquad}, g, h)$
	$= (b, a, i, c, d, e, f, g, h)$

Figure 6.3 Crossover operators PMX and OX.

Both, PMX as well as OX, assemble one offspring permutation π_o from two parental permutations π_1 and π_2. For both techniques a matching section (underlined) is determined randomly in π_1 at first. Afterwards PMX copies the elements of π_2 which do not belong to the matching section in two steps to π_o. Those elements are passed first to the offspring which can keep their position before the rest is copied in some way that creates a valid permutation. Finally the matching section is implanted at the same position in π_o it has in π_1. In contrast, OX copies those elements of π_2 to π_o at once that do not belong to the matching section. These elements are pushed together such that the matching section can be implanted in the offspring right after that element which connects the first element of the matching section in π_2 (in the example this connection is 'i, c'). Here the idea is to properly pass on the relative ordering at the cut of the matching section from π_2 to π_o.

Fig. 6.3 shows that PMX and OX implement the fraction passed from π_1 without modification in the offspring. The fraction passed from π_2 still gets little distorted in order to ensure the generation of valid permutations. Consequently the size of the matching section has to be smaller than half the permutation length. Otherwise the proportion of information taken from π_1 would predominate in the long run. Usually the size of the matching section is drawn within the range of 1/3 and 1/2 of the total length.

Additionally we deal with the *precedence-preservative crossover* (PPX) introduced before by Bierwirth et al. (1996). This operator passes on precedence relations among the elements of two parental permutations to one offspring at the same rate. PPX works as follows. A vector at the length of the permutation is randomly generated over the alphabet $\{1, 2\}$. It serves as a preference list defining the order in which elements are successively drawn from π_1 and π_2[9]. PPX initializes an empty offspring. The leftmost element in one of the two parents is selected in accordance to the preference list. After an element has been selected it is deleted in both parents and appended to the offspring. This step is repeated until both parents are empty.

parent 1	$\pi_1 = (a, b, c, d, e, f, g, h, i)$
parent 2	$\pi_2 = (f, b, e, a, i, c, g, d, h)$
preference list	1, 2, 2, 1, 1, 1, 2, 1, 2
	$\pi_o = (a)$
	$= (a, f)$
	$= (a, f, b)$
	...
	$= (a, f, b, c, d, e, i, g, h)$

Figure 6.4 Crossover using the Precedence preservative operator PPX.

The three operators end up with a valid offspring permutation. We expect PMX to pass on absolute ordering most reliable. OX probably preserves relative ordering best. However, no operator completely avoids the introduction of some new absolute and relative ordering. Only PPX results in a proper preservation of the global precedence observed in parental permutations.

In order to examine whether the operators meet our expectations we apply them to 1,000 pairs of random permutations of length $n = 100$ each. The normalized distances of absolute ordering, relative ordering and global precedence are computed for the parents and their offspring respectively. This measure, averaged over the 1,000 samples, is reported in Tab. 6.2.

Let us have a look at the distances measured for arbitrary permutations π_1 and π_2 first. Regarding absolute and relative ordering we observe maximal distances (≈ 1.0), which means that random permutations hardly share positional assignment and adjacency of their elements. Furthermore, on average one half of the precedence relations are common in random permutations which exactly meets the expected value under a uniform distribution.

The observed distances between π_1, π_2, and the offspring π_o are smaller in all cases. This indicates that learning by crossover takes place in terms of passing a considerable amount of information. Strong

[9] In the context of vehicle routing problems a similar operator was independently introduced by Blanton and Wainwright (1993) under the name *merge crossover*. Different than PPX, merge crossover incorporates a "merging list" depending on problem data instead of random preferences.

Table 6.2. The averaged normalized distance between two arbitrary parent permutations and between the corresponding parent and offspring permutations.

crossover technique	$d_a(\pi_1, \pi_2)$	$d_a(\pi_1, \pi_o)$	$d_a(\pi_2, \pi_o)$
PMX		0.62	0.61
PPX	1.00	0.99	0.98
OX		0.99	0.97
	$d_r(\pi_1, \pi_2)$	$d_r(\pi_1, \pi_o)$	$d_r(\pi_2, \pi_o)$
PMX		0.62	0.75
PPX	0.99	0.82	0.81
OX		0.61	0.61
	$d_p(\pi_1, \pi_2)$	$d_p(\pi_1, \pi_o)$	$d_p(\pi_2, \pi_o)$
PMX		0.29	0.36
PPX	0.50	0.25	0.25
OX		0.45	0.23

quantitative differences between the operators can yet be found. Regarding absolute ordering merely PMX results in a clear similarity of parents and offspring. Since π_1 and π_2 share nearly no position-based information, a distance value of $d_a \approx 0.6$ from π_o to both parents means that the offspring carries about 40% of this kind of information from the parents each. Concerning relative ordering PMX also works effectively but it cannot balance the parental bias. It obviously favors parent π_1 which provides the matching section. OX works better since it allows the parents to pass on information in equal shares. Looking at global precedence things change again. Only PPX treats both input permutations on an equal basis. At the same time PPX perfectly preserves precedence information which is verified by adding $d_p(\pi_1, \pi_o)$ and $d_p(\pi_2, \pi_o)$. This sum corresponds to the distance of random permutations. Using a mapping section conspicuously favors π_1 or π_2 in PMX and OX respectively. Like for mutation considered before, every crossover operator dominates one ordering categories and cut off worst in another.

Based on the above design guideline for crossover operators a certain technique appears the more useful the more reliable it respects the ordering category emphasized by a problem class under consideration[10]. This reasonable claim is confirmed by countless studies dealing with permuta-

[10] While adaptation progresses, the available knowledge becomes increasingly similar. If the EA starts converging, further exploration is hindered by too preservative crossover operations. A rough technique might become advantageous. The same effect is also reached by increasing the mutation frequency, which allows a careful convergence control at the same time.

tion problems, but it is probably most evident for the TSP. Here relative ordering is emphasized exclusively and therefore it cannot surprise why advanced operators which control the parent-offspring distance edge for edge are very successful, see Freisleben and Merz (1996).

4. SUMMARY

In this chapter we have focussed on basic operations of logistics systems such as sequencing, scheduling, grouping, or packing of goods. Corresponding decision problems in practice usually nest inside one another resulting in great challenge for optimization. A promising idea to cope with this complexity relies on agent technology. Using adaptive agents being responsible for some subtask appears attractive because the reconciliation of mutually dependent decisions might develop from adaptation. On this background we have reflected different ways of representing logistics domain knowledge for adaptive agents.

The problem representation system focuses on two components, the learning operators and the encoding-decoding system. For combinatorial problems a good deal of research concentrates on permutation encoding because this often allows to describe solutions quite naturally. However, the different properties of permutations can be stressed to fairly different extents. The evident importance of relative ordering in sequencing and of position in assignment is due to the fact that these problems are usually unconstrained in the permutation scheme. Turning to mixed-type problems a permutation covers the implicit constraints of a problem which remain invisible until the permutation has been decoded. Therefore it is often not clear in advance whether an operator should maintain relative or absolute ordering. The most pervasive property of a permutation is expressed by the global precedence of its elements. A new crossover operator has been designed which is able to perfectly preserve this fine-grained information.

Chapter 7

ADAPTIVE SCHEDULING

In the area of industrial production planning, scheduling problems have been subject of intensive research over the last 50 years. The common focus of scheduling is on the efficient allocation of one or more *resources* to *activities* over time. Due to its convenience, the terminology used in production is also adopted for scheduling in other fields, e.g. in transportation, public traffic systems, power plants, to mention just a few. We follow this line and refer to a *job* as a complex consisting of one or more activities and to a *machine* as a resource that can perform one activity at a time. Among the variety of different scheduling problems the classical *job shop problem* is the most studied one by academic research. It can be doubted, however, whether the problem owes its reputation from an outstanding importance in shop-floor management. Viewing the job shop problem as a benchmark for the comparison of scheduling algorithms is probably more suitable. For this purpose the job shop problem is undoubtedly an ideal candidate because it generates an enormous complexity from few input data by including at least some features of the real-world.

In order to clarify the intractability of schedule optimization the classical job shop problem is presented in the first section. A highly expressive graph representation is introduced for this standard model. To assess the suitability of local-search based algorithms for job shop scheduling, we next analyze the landscapes of such problems. Central components of local search, including those of evolutionary methods, are reviewed in the second section. We investigate the prospects and limits of EAs for scheduling by computational experiments and by a comparison with other promising heuristics. Conclusions regarding the impact of EAs on real-world scheduling problems are finally discussed.

1. CLASSICAL JOB SHOP SCHEDULING

In the field of combinatorial optimization the *job shop problem* (JSP) is known as one of the most intractable *NP*-hard problems since long. A bulk of literature on this problem has accumulated over the years including exact methods as well as heuristic approaches, see Błażewicz et al. (1996). In this section the JSP is introduced informally at first. Afterwards a mathematical formulation based on a graph notation is given and illustrated by a small example. Then we present ten large-scale benchmark instances. Finally, the structure of JSP fitness landscapes is investigated by a configuration space analysis for the benchmark problems at hand. The results gained give us a clue concerning the potentials of EAs for scheduling.

1.1 PROBLEM DESCRIPTION

The JSP is stated in short as follows. Consider n jobs consisting of a fixed number of activities, called *operations* in this context. Every operation allocates one of m dedicated machines for a prescribed processing time. For each job a machine routing is defined which expresses a technological order of processing the operations of that job. The objective pursued is to find a job sequence for every machine such that the resulting machine schedule is feasible and the time span needed to make all jobs (*makespan*) takes a minimum. Several assumptions are made implicitly.

1. There is only one of each type of machine.
2. No machine may process more than one operation at a time.
3. No two operations of one job may be processed simultaneously.
4. Processing an operation may not be interrupted (no preemption).
5. No job is processed twice on the same machine.
6. Jobs may be started at any time, i.e. no release times exist.
7. Jobs may be finished at any time, i.e. no due dates exist.

Next to mixed-integer models, the *disjunctive graph formulation* presented by Adams et al. (1988) has become popular for schedule optimization approaches based on principles of local search and adaptation.

1.2 DISJUNCTIVE GRAPH FORMULATION

In the disjunctive graph formulation of the JSP two additional operations are introduced which represent the begin and the end of the entire production program. Let V denote the set of all operations of the n jobs including the two mentioned dummy operations. Furthermore, let A denote the set of ordered pairs of operations representing the technological precedence relations among the operations of a job. Thus, if $(v, w) \in A$ with $v, w \in V$, this means that operations v and w belong to the same job and that w has to be processed as a successor of v. Let M denote the set of the m machines and let E_j denote the set of all pairs of operations to be processed on the same machine $j \in M$. Notice that processing operations from E_j may not overlap with respect to assumption 2 made above.

For each operation $v \in V$ a fixed processing time p_v is prescribed. The processing times of the two dummy operations are set to zero. The problem is to determine starting times t_v for all operations $v \in V$ such that the completion time of the latest job is minimized. This is as to minimize the starting time of the ending operation, i.e. $\min t_e$ subject to

$$t_v \geq 0 \qquad \text{for all } v \in V, \tag{7.1}$$

$$t_w - t_v \geq p_v \qquad \text{for all } (v, w) \in A, \tag{7.2}$$

$$t_w - t_v \geq p_v \text{ or } t_v - t_w \geq p_w \qquad \text{for all } (v, w) \in E_j,\ j \in M. \tag{7.3}$$

Conditions (7.1) ensure that all operations are processed while conditions (7.2) effect that the sequence of processing operations in a job corresponds to the prescribed machine routing of that job. Finally conditions (7.3) care that no two operations are processed simultaneously by the same machine. A complete table of operation starting-times t_v for which conditions (7.1) to (7.3) hold is called a *feasible schedule*.

In the associated graph formulation of the problem there is a node for each operation involved in a problem. The dummy operations represent the source and the sink in the graph. Every two consecutive operations of a job define a *conjunctive* arc $(v, w) \in A$ in the graph. For each pair of operations that require the same machine j there is a pair of *disjunctive* arcs $(v, w), (w, v) \in E_j$ with opposite directions in the graph. Let $E = E_1 \cup E_2 \ldots \cup E_m$ denote the set of all disjunctive arcs. Finally, every arc $(v, w) \in A \cup E$ is labeled by a weight $p_v \in P$ corresponding to the processing time of operation $v \in V$.

All this results in a weighted directed graph $G = (V, A \cup E, P)$ with nodes in V, arcs in $A \cup E$, and weights in P. Because the JSP requires to determine an operation sequence for the machines, this means to select one arc among the disjunctive arc pairs in G such that the resulting graph G' is acyclic. Otherwise, if G' has a cycle, there are precedence conflicts between operations violating (7.1)-(7.3) which prevent building a feasible schedule.

For an acyclic graph we can determine the maximum-weighted path connecting the source node and the sink node. Obviously, the length of this so called *critical path* is equal to the starting time t_e we seek to minimize. In other words, the JSP is stated as to make a decision for all disjunctive arc pairs such that (i) the resulting graph gets acyclic and (ii) the length of a critical path in this graph gets minimized.

1.3 AN ILLUSTRATING EXAMPLE

We consider a JSP instance with $n = 3$ jobs and $m = 3$ machines. The problem data prescribes specific machine routings for the jobs and different processing times for the operations. Notice that operations 1, 2, and 3 belong to job 1, operations 4, 5, and 6 belong to job 2, and so on.

Table 7.1. Data of a 3-job and 3-machine job shop scheduling problem.

job	*machine routing*	*operation processing times*
1	M1, M2, M3	$p_1 = 3$, $p_2 = 3$, $p_3 = 2$
2	M2, M3, M1	$p_4 = 2$, $p_5 = 3$, $p_6 = 3$
3	M2, M1, M3	$p_7 = 4$, $p_8 = 3$, $p_9 = 1$

Formulating the problem by a disjunctive graph G is shown in Fig. 7.1. The conjunctive arcs (representing the machine routings) are shown solid whereas the pairs of disjunctive arcs (connecting operations to be processed on the same machine) are shown dashed. According to the operation processing-times, the outgoing arcs of each node are equally weighted. A schedule results by determining operation sequences for each machine. As one easily verifies, only a fraction of the possible $n!^m = 216$ combinations of machine sequences can be realized consistently. Sequencing for instance operation 8 prior to operation 1 on M1 and operation 2 prior to operation 7 on M2 leads to a deadlock because neither job 1 nor job 3 can be started without violating the prescribed machine routing.

In the disjunctive graph model, machine sequences result from selecting one arc among every disjunctive arc pair. Consistency is achieved by avoiding cycles. Notice that in the above sketch constellation a cycle of operations occurs, namely (1,2,7,8,1). Presupposed a consistent selection has been made for the disjunctive arc pairs in G, unique machine sequences are fixed by means of the specified precedence among operations.

Fig. 7.2 shows one possible solution for the problem under consideration. The dashed arcs of this graph G' indicate the sequences of

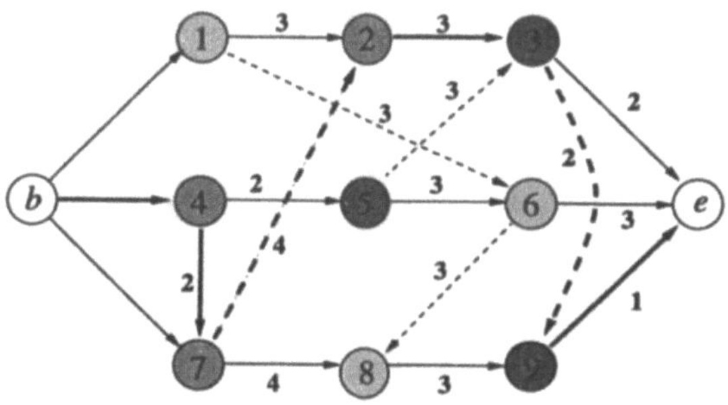

Figure 7.1
Disjunctive graph for the example instance.

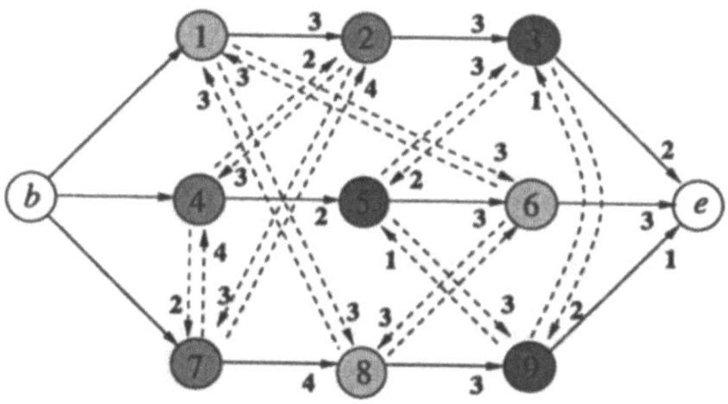

Figure 7.2
Schedule representation in the graph model.

processing operations on the three machines. While computing a critical path from the source node to the sink node in G', a complete table of starting times for the operations is built up[1]. Summing up the weights along this critical path (shown by bold figured arcs) leads to a value of 12 time units which expresses the required makespan of the corresponding schedule.

Notice that none of the operations belonging the critical path can be delayed a little without deteriorating the schedule at the same time (i.e. increasing the makespan). Every operation with this property is referred to as a *critical operation.* In general, any critical path defines a sequence of critical operations connecting the source and the sink node[2]. A subsequence of critical operations on the same machine is called a *critical block.*

A more vivid impression of schedules is obtained from so called *Gantt charts,* see Fig. 7.3. Here each operation is depicted as a rectangular whose length corresponds the prescribed processing time. The operation starting-times as well as the required makespan can be taken directly from the abscissa. Orienting a Gantt chart according to machines or

[1]For a detailed description of an efficient implementation of this longest-path algorithm see Mattfeld (1996).

[2]Notice that operations 5 and 6 are also critical which is why there exists a further critical path in G' along the nodes 4, 5, 6, 8 and 9.

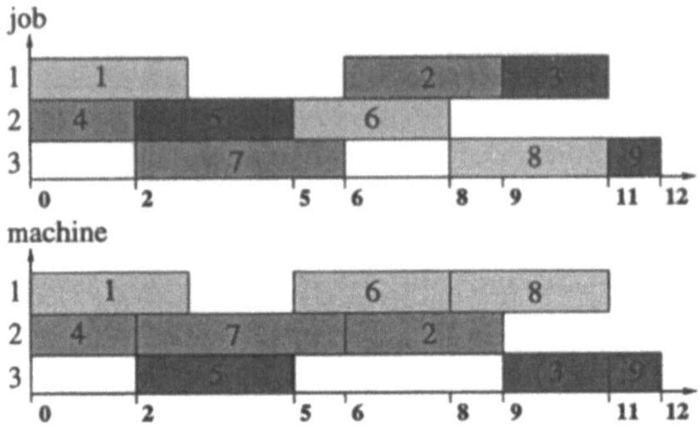

Figure 7.3
Schedule representation by Gantt charts.

jobs either visualizes intermediate job waiting-times or machine idle-times. This interface is extremely useful to reveal possible ways for improving schedules in a person-machine interaction.

1.4 BENCHMARK PROBLEMS

In a job shop each job has to pass the machines in a specific technological order, referred to as its machine routing. For benchmark problems the machine routings are often generated as uniformly distributed machine sequences. Another way to generate these constraints is to partition the set of machines M at first and to generate random sequences for each machine partition afterwards. Since all jobs have to pass the machines of one partition before the machines of the next partition are taken into account, a *work-flow* is introduced to the job shop. This technique is generalized by use of a variable q $(1 \leq q \leq m)$ denoting that the set of machines is q-partitioned, i.e. $M = M_1 \cup M_2 \ldots \cup M_q$ and $M_i \cap M_j = \emptyset$ for all $1 \leq i < j \leq q$. In order to achieve that the q partitions are of equal size the consideration of q is often restricted to integer based dividers of m. The generation of machine routings is carried out now by subsequently generating a random sequence of m/q machines within each of the q partitions. Due to this construction the machine routings are generated independently if and only if $q = 1$. For $q > 1$ a work-flow is introduced in the job shop and for $q = m$ the job shop has turned into a flow shop.

For computational investigations on the JSP we refer to a collection of ten benchmark problems generated by Storer et al. (1992)[3]. Although the problems consist of the same number of jobs ($n=50$) and machines ($m=10$) each, five instances are very hard to solve while the others are relatively easy. The hard problem instances are generated with $q=2$

[3]The entire testbed consist of twenty problems of which the larger ten ones, referred to as 10x50e1-10x50e5 and 10x50d1-10x50d5 are considered here.

Table 7.2. A comparative testbed of easy and hard to solve JSP benchmarks. Columns ϵ_{rand} and ϵ_{adapt} refer to the observed mean relative error of random schedules and local-optimal schedules against the best-known solutions.

problem $(m \times n)$	*best-known*	ϵ_{rand}	ϵ_{adapt}
	2924	0.55	0.13
	2794	0.58	0.18
"easy" (10×50)	2852	0.53	0.11
	2843	0.68	0.25
	2823	0.54	0.12
	average	0.58	0.16
	3047	1.14	0.65
	3012	1.16	0.66
"hard" (10×50)	3122	1.12	0.66
	2968	1.17	0.67
	2924	1.16	0.69
	average	1.15	0.67

and the easy problem instances with $q = 1$ respectively. The smaller the partitions of M gets the more jobs attempt to allocate the same machines at the same time. This makes a problem difficult in turn. In reverse, for $q=1$, the machine routings are uniformly distributed which reduces the possible conflicts among jobs[4].

For each test-problems the best-known makespan is reported in the column "known" of Tab. 7.2. For the hard problems no. 1 and 4, these schedules have been found by Balas and Vazacopoulos (1998) while the solutions of the remaining problems no. 2,3 and 5 have been recently improved by Steinhöfel et al. (1998). Only the hard problem no. 4 is proved to be solved to optimality while, of course, all easy problems are.

To get a first glimpse of the testbed, two pools of solutions are generated for the problems each. One pool (*rand*) consists of 1,000 randomly generated schedules. The other pool (*adapt*) consists of a corresponding number of local-optimal schedules which are obtained by an iterative improvement algorithm, starting from the random schedule and applying the best-improvement strategy. The further columns in the table give the mean relative error ϵ in both solution pools for each problem instance. Thereby the relative error of a single schedule s is calculated from its makespan f_s and the best-known makespan by $\epsilon_s = f_s/f_{known} - 1$.

[4]Especially if n is significantly larger than m, the $q=1$ technique induces problem easiness, compare Taillard (1993).

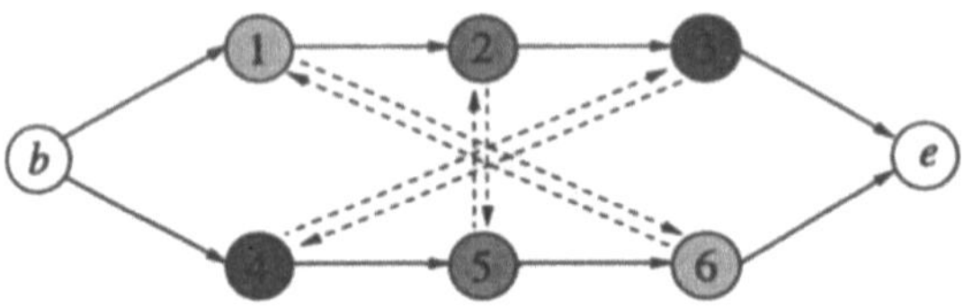

Figure 7.4 Disjunctive graph formulation.

Solutions of the hard as well as of the easy instances show apparent similarities regarding the mean relative error. E.g. random schedules of hard instances have an error of 1.15 versus 0.67 for the corresponding adapted schedules. For easy instances this relation decreases from 0.58 to 0.16. Therefore all twenty pools are divided into four classes by combining the attributes *hard/easy* and *adapt/rand* respectively. In the following these classes are subject to statistical computations for which results are averaged over all solution pools belonging to the same class.

1.5 THE FITNESS LANDSCAPE

Although the JSP has received much attention in recent years, there are only a few results on the structure of its fitness landscape[5]. By comparing characteristic properties of landscapes of different JSP instances among each other and by comparing them with the characteristic properties of landscapes of other combinatorial problems, we aim to attain an understanding of the intractability of the JSP.

Recall from Chap. 4(2.1) that introducing a landscape for a combinatorial optimization problem requires a suitable problem representation and, on this basis, a neighbourhood definition. The smallest possible modification of a schedule reverts a single disjunctive arc. In this way a schedule $s_1 \in S$ is moved into a neighbor s_2 by swapping the precedence relation of two adjacent operations on one machine without violating the prescribed machine routings, which would produce a cycle in the corresponding graph. In reference to this *basic move*, a set of peaks $P \subset S$ is clearly defined in the fitness landscape of a problem instance. These peaks are accessed from arbitrary schedules by iteratively applying improving basic moves until no further improvement can be gained.

Example 7.1 Consider a small JSP instance consisting of $n = 2$ jobs and of $m = 3$ machines. The operation processing times are ignored for simplicity. The jobs have different machine routings. Job 1 consecutively passes machine M1, M2, and M3, whereas job 2 passes them in inverse direction (M3, M2, M1). We denote the successive operations of job 1 as

[5] cf. Mattfeld (1996); Mattfeld et al. (1999).

s	*machine sequences*		
	M1	M2	M3
111	1,6	2,5	3,4
110	1,6	2,5	4,3
101	–	–	–
100	1,6	5,2	4,3
011	–	–	–
010	–	–	–
001	–	–	–
000	6,1	5,2	4,3

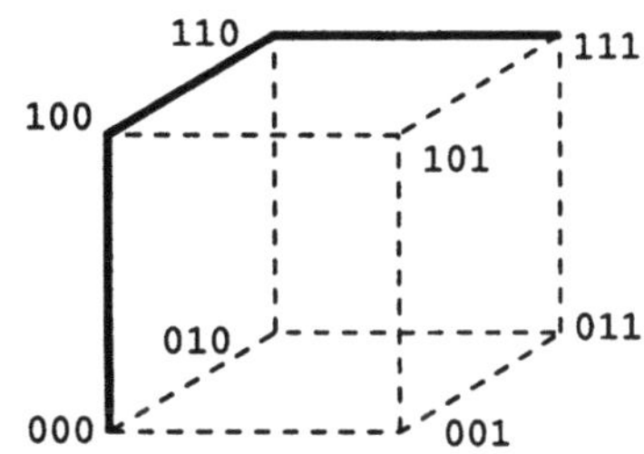

Figure 7.5. Binary vs. natural representation of schedules for a small JSP. The three bits of a string consecutively encode the operation sequences on the three machines.

1, 2, 3 and the operations of job 2 as 4, 5, 6 respectively. The disjunctive graph corresponding to this problem is sketched in Fig. 7.4.

A binary representation of the problem results from encoding the orientation of all disjunctive arc pairs. Let bit "1" denote that the downward-oriented arc is selected from a particular disjunctive arc pair in the graph of Fig. 7.4. Altogether only one half of the 2^3 possible selections lead to feasible schedules. Notice in particular the high probability of conflicts on M2. The structure of the corresponding search space S is shown by the bold edges of the boolean cube in Fig. 7.5. Like for the TSP, a basic neighborhood move corresponds to a single bit-flip, compare Chap. 4(2.4). Unlike the TSP, the feasible solutions of a JSP are not produced with identical probability anymore. Since the strings 111 and 000 have more infeasible neighbors than 110 and 100 they are prefered by random sampling.

The dimensionality of the landscape and its cardinality are $D_s = 1.5$ and $C_s = 4$ respectively. According to equation (4.3) we estimate the landscape's modality by $M = 4/2.5 = 1.6$. Already by taking a look at this small example it is verified that JSP landscapes can be more rugged than TSP landscapes. There, a modality of $M = 2$ has been observed for a problem of corresponding size. This indicates that the degree of epistatic interaction for the JSP goes far beyond the corresponding level of the TSP, although we cannot demonstrate this analytically in terms of the NK-model. ∎

For an empirical classification of JSP landscapes we refer to the ten benchmark instances of Tab. 7.2. In the first step we concentrate on the change of entropy which is observed when schedule adaptation takes place by means of iterative improvements. For this purpose the en-

Table 7.3. The change of entropy ΔE between randomly generated and adapted schedules observed in the JSP testbed. Corresponding results are provided for the QAP and the TSP.

problem type	E_{rand}	E_{adapt}	ΔE
JSP (easy)	0.434	0.429	0.005
JSP (hard)	0.678	0.675	0.003
QAP (uniform)	1.00	0.97	0.03
QAP (Elshafei)	1.00	0.80	0.20
TSP (symmetric)	1.00	0.32	0.68

tropy is calculated by formula (4.1) for the pools of schedules addressing the four attribute classes introduced before. Tab. 7.3 shows the values obtained together with the change of entropy observed. Recall from Chap. 4(2.3) that an entropy close to one is usually expected for a pool of randomly generated solutions. For the JSP we measure a significantly smaller entropy which confirms the observation made above that the precedence among operations is not uniformly distributed in random sampling. The fact that the entropy of hard instances is clearly larger than the entropy of easy instances cannot surprise. Less conflicting machine routings of the easy problems allow a definite fixation for a large number of precedence relations. In spite of these differences, easy and hard problems show nearly the same change of entropy under adaptation. Since ΔE is always very low, we conclude that local optima are widely spread in the landscapes, independent of whether a certain JSP is more or less intractable.

For other combinatorial problems completely different results are obtained. For the QAP random sampling produces all possible solutions with identical probability. As reported by Taillard (1995), adaptation hardly reduces the entropy at least for instances with uniformly distributed distance matrices. For a QAP with more structure like the famous real-world problem of Elshafei, the entropy yet decreases to the extent of 20%. For symmetric TSPs the situation is different again. Here adaptation reduces the entropy drastically by about 68%.

Since entropy measures cannot explain the different intractability of JSP instances we focus on the force of attraction of their landscapes next. Tab. 7.4 shows the mean Hamming distance $\overline{r}_{rand}$ between random schedules (measured on the basis of a binary mapping) and the average length of adaptive walks $\overline{r}_{adapt}$ in each of the four attribute classes. The larger basins of attraction for the easy instances, expressed by Λ, confirms the prevailing opinion that problem easiness and the smoothness

Table 7.4. The expected force of attraction Λ of local-optimal schedules and the correlation length h^* observed for large walks in the fitness landscapes of the testbed.

JSP problems	$\bar{r}_{rand}$	$\bar{r}_{adapt}$	Λ	h^*
"easy"	2375	87.6	0.037	270
"hard"	3860	103.3	0.027	980

of fitness landscapes are related to a certain extent. At first glimpse a difference of one percent in both values seems not to be much. But in contrast we argue that it clearly enhances the chance of adaptation to succeed. Recall from Chap. 4(2.3) that e.g. for $\Lambda = 0.1$ already every tenth solution can reach the global optimum via adaptive walks with a probability larger than zero.

The different structures of JSP landscapes become even more obvious by taking a look at the fitness correlation observed for large random walks on the different landscapes. Since random sampling of schedules is not unbiased, landscapes of the JSP are not statistically isotropic. Nevertheless, we have measured the overall correlation for the five benchmarks in both problem classes by performing ten walks of 100,000 steps each. The autocorrelation is calculated using equation (4.5) and averaged over the ten walks. The assumed correlation length h^* is shown in the last column of Tab. 7.4. Although these measurements may be distorted through a bias in the landscapes, they indicate an immanent difference between easy and difficult to solve problems. Landscapes in the class of the difficult instances show a much planer surface than landscapes of the easier instances. The latter are obviously richer in structure which is expected as a valuable references to problem easiness.

2. LOCAL-SEARCH ALGORITHMS

The great interest in scheduling has made the JSP to a popular challenge for local-search algorithms. For an investigation of these approaches we distinguish between ***neighborhood search*** and ***adaptive search.*** The characteristics of both kinds of search are listed in Tab. 7.5 according to the classification of local search algorithms in Fig. 4.7.

In this section the problem-specific components of neighborhood search and adaptive search are sketched for the JSP. Afterwards we compare different adaptive search strategies by a computational study. In order to reveal the potentials of adaptive search for scheduling, these results are finally considered in the context of the performance of other local-search algorithms.

Table 7.5. A simple differentiation of local-search algorithms.

	neighborhood search	*adaptive search*
current solutions	point-based	population-based
exploration	single neighborhood moves	long recombination jumps
guidance	stochastic acceptance	statistical selection
variants	Sim. Anneal., Tabu Search	Genetic Algorithms, etc.
hybrid variants	e.g. backtracking	e.g. genetic local-search

2.1 NEIGHBORHOOD SEARCH

The convincing advantage of the disjunctive graph formulation is to open the JSP for iterative improvement algorithms. A necessary condition to improve a given schedule is to start at least one critical operation earlier than in the original schedule. Whether this is possible without violation of constraints can be explored by a neighborhood search which reorders the operations in a critical block.

Based on this idea, van Laarhoven et al. (1992) have introduced the *critical-transpose neighborhood.* It is defined as follows. Let S denote the set of all feasible schedules of a problem instance. A neighbor s_2 is derived from a schedule $s_1 \in S$ by swapping two adjacent operations in the same critical block. Any schedule obtained by this technique is feasible, i.e. $s_2 \in S$ always holds. This is verified in Fig. 7.6 which illustrates that swapping of adjacent operations in a critical block cannot produce a cycle. With respect to this argument the critical-transpose neighborhood is actually a restriction of the basic JSP neighborhood introduced in the previous section.

A useful restriction of the critical-transpose neighborhood, already proposed by Matsuo et al. (1988), is due to the observation that no improvement of s_1 can result from a reordering, if the first operation and the last operation of a block remain untouched in s_2. The *critical-end neighborhood* is a restriction of the critical-transpose neighborhood where only the first and last operations of a block can be swapped.

Different neighborhoods have different properties which makes them more or less suitable regarding the quality and the speed at which a local-search algorithm proceeds. Concerning the runtime demand the critical-end neighborhood appears highly attractive for two reasons. First, the size of the neighborhood (i.e. the number of neighbors of a single solution) is relatively small. In order to explore the neighborhood of a schedule completely, only a few other schedules have to be examined. Second, the examination of neighboring schedules can take advantage

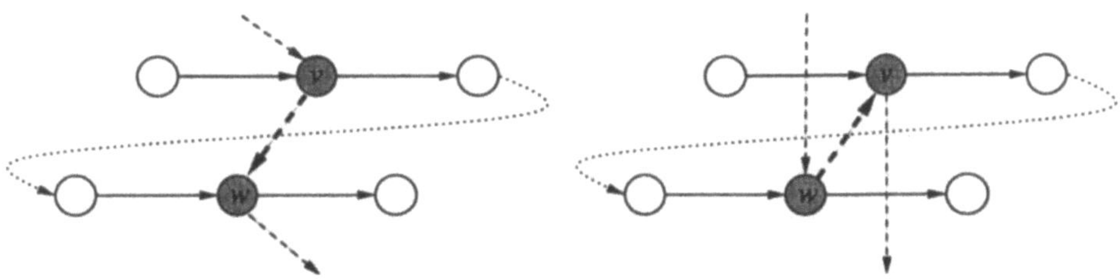

Figure 7.6. The critical-transpose neighborhood: Consider any two adjacent operations v and w in a critical block. According to the definition these operations are processed without delay immediately after each other on the same machine (shown on the left). A path connecting the job successor of v (the subsequent operation of the same job) and the job predecessor of w (the previous operation of the same job), which is suggested by the doted arc, cannot exist, because the length of this path certainly would exceed the weight of the critical arc (v, w). Consequently, inverting this disjunctive arc into (w, v) cannot produce a cycle (shown on the right) because this definitely requires the doted connection which does not exist.

from an efficient lower-bound calculation which provides a good estimate on the success of a particular neighborhood move, see Taillard (1994). Although the JSP does not allow an update of the objective function value as rapid as e.g. the TSP or the QAP[6], a complete recalculation of a neighboring schedule can be avoided in many cases.

An essential drawback of too small neighborhoods is that the final solution may be of poor quality. Therefore Dell' Amico and Trubian (1993) have extended the simple block-modification approach by sophisticated reordering techniques which are more effective than swapping adjacent operations. As a result, the size of the neighborhoods increases and the quality of the schedules which can be obtained improves significantly. Further, even more efficient neighborhoods have been suggested by Balas and Vazacopoulos (1998) and Mastrolilli (1998) for the JSP[7].

Various approaches in Simulated Annealing and Tabu Search have been proposed for schedule optimization throughout the last decade. Without claiming completeness we mention the work of Matsuo et al. (1988); van Laarhoven et al. (1992); Dell' Amico and Trubian (1993); Taillard (1994) and Steinhöfel et al. (1998). Other procedures performing a multilevel search by resembling limited tabu walks with backtracking have been developed by Nowicki and Smutnicki (1996) and Balas and Vazacopoulos (1998). These hybrid variants of neighborhood search turned out to be especially successful for the classical JSP.

[6] compare equations (6.7) and (6.8).

[7] For properties of neighborhoods extending the critical-transpose one see Vaessens (1995).

2.2 ADAPTIVE SEARCH

Since the publication of Davis (1985) early paper on EAs for scheduling problems, numerous adaptive approaches have been reported in literature. For a survey see Portmann (1996) or Tsujimura et al. (1997). Most of the algorithms proposed actually point to GAs and its order-based cousins. This section gives a rough overview on some concepts which are widely used in evolutionary computation. Particular attention is paid to the problem encoding and to the decoding of schedules.

ENCODING AND RECOMBINATION

In the graph formulation of the JSP, disjunctive arcs are used in order to express possible decisions about the precedence of operations to be processed on the same machines. Let $\text{prec}_j(k,l) = 1$ if job k precedes (not necessary directly) job l on the j-th machine and $\text{prec}_j(k,l) = 0$ otherwise. Hence, if all n jobs have to pass all m machines there is a total of $m \cdot n(n-1)/2$ precedence relations among the operations involved. This can be used to introduce a fixed-length binary encoding for the JSP. An illustration for the schedule in Fig. 7.2 is given below.

machine j	1			2			3		
operation k	1	1	6	2	2	4	3	3	5
operation l	6	8	8	4	7	7	5	9	9
$\text{prec}_j(k,l)$	1	1	1	0	0	1	0	1	1

Figure 7.7 Binary representation of machine sequences.

This unique binary mapping has been used by Nakano and Yamada (1991) in order to apply a conventional GA to the JSP. Unfortunately, standard recombination operators produce infeasible schedules almost always. For this reason a repair mechanism has to be employed which effects that (i) valid machine sequences can be decoded from arbitrary binary strings and that (ii) the corresponding graph does not become cyclic. However, the binary encoding hardly respects the combinatorial nature of the JSP which may explain why the computational results achieved are rather poor.

In order to rely on the combinatorial properties of the JSP as a multi-sequencing problem, it is formulated according to the general encoding scheme (6.6). A set of n objects (jobs) has to be sequenced on m resources (machines). Each object consists of a set of m activities. The temporal order of resource allocations is restricted by dependencies among the activities. Find a permutation

$$\pi = \{\underbrace{\pi_1, \ldots, \pi_n}_{\text{machine } 1}, \underbrace{\pi_{n+1}, \ldots, \pi_{2n}}_{\text{machine } 2}, \ldots, \underbrace{\pi_{(m-1)n+1}, \ldots, \pi_{mn}}_{\text{machine } m}\} \tag{7.4}$$

of activities π_i which is m-partitioned by the multiples of n such that the time-span needed to perform all activities is minimized. From this it is clear that the JSP belongs to case (iv) of the problem classification in Chap. 6(2.1).

Interpreting permutations as multi-sequences of operations can still result in infeasible schedules. Consider the example problem in Tab. 7.1 again. Assumed that none of the operations 1, 4, and 7 is placed in the first position of a partition of π, no operation can be scheduled for any machine. The representation of infeasible schedules is avoided by slightly modifying the permutation scheme as independently proposed by Fang et al. (1993) and Bierwirth (1995). Instead of using a partitioned permutation of operations we change to an unpartitioned *permutation with repetition* of jobs. For each job an identifier occurs as often in the permutation as there are operations belonging to this job. While scanning such a permutation from left to right, the i-th occurrence of a job refers to the i-th operation in the machine routing of this job. In this way one avoids scheduling operations whose job predecessor has not been scheduled before. Fig. 7.8 shows a permutation from which the schedule in Fig. 7.3 can be assembled.

permutation w. repetition	1	2	2	3	1	2	3	1	3
index of occurrence	1	1	2	1	2	3	2	3	3
points to operations	1	4	5	7	2	6	8	3	9
partitioned permutation	1	6	8	4	7	2	5	3	9

Figure 7.8 Transformation of permutations.

Notice that the machine sequences are faded out in this encoding. Actually, machine sequences do not evolve until a permutation has been transformed with respect to the machine routings prescribed. Afterwards a corresponding m-partitioned permutation can be established by moving operations belonging to the same machine into one partition while keeping their mutual precedence relations unchanged. In result, all operations can be scheduled one by one without running into a deadlock. However, complicated operators for recombination have to be used in order to keep the particular repetition structure of a problem, see Bierwirth (1995) and Mattfeld (1996).

Representing *priorities of operations* instead of operation sequences, as proposed by Della Croce et al. (1995), helps to counter this drawback. Recall that operations become available if their job predecessor is completed. Whenever a machine becomes idle, an available operation is chosen according to a priority list of operations and scheduled next. Basically a priority list is a permutation again, with the difference that it is interpreted as a precedence matrix, compare Example 6.1. In

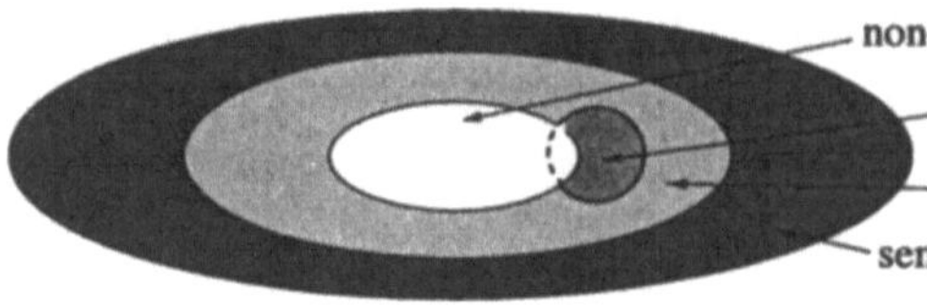

Figure 7.9 Relationships of schedule properties.

this way priorities are prescribed between any two contained operations. This always creates a feasible schedule. Since the priority representation relies on ordinary permutations it basically allows to use any of the crossover operators of Chap. 6(3.3). The encoding, however, predominantly stresses the ordering by global precedence and therefore PPX is the operator of best choice.

It is worthwhile to mention that many other encodings have been proposed for the JSP, see Bruns (1993); Dorndorf and Pesch (1995) or Rixen (1997). We do not discuss these approaches as they basically rely on the decoding strategies presented below.

SCHEDULE DECODING

Any encoding of schedules can be decoded into *semi-active*, *active*, or *non-delay* schedules. The three properties of schedules are achieved by increasingly incorporating domain knowledge which continuously reduces the space of accessible schedules, see Fig. 7.9. A description of these schedule properties is presented below in reference to a decoding technique using the priority representation of operations.

Semi-active schedules are built by assigning operations to their earliest possible starting time. This means that any operation is immediately started at the maximum completion time for the previous operation of the same job or of the same machine. The permutation π serves as a look-up array in order to solve possible conflicts among the operations available at a certain point in time. Such kind of procedure reads as follows.

1. Build the set B consisting of the beginning operations of all jobs (dummy operations are not considered).
2. Select operation o^* from B which occurs leftmost in π and delete it from B, i.e. $B := B \backslash \{o^*\}$.
3. Schedule o^* for processing at the earliest possible starting time.
4. If a job successor o^*_s of o^* exists, insert it into B, i.e. $B := B \cup \{o^*_s\}$.
5. If $B \neq \emptyset$ goto Step 2, else terminate.

Notice that this procedure applies the FCFS principle if operations of π are scheduled consecutively without running into a deadlock.

Active schedules are produced by modifying step 2 in the above procedure which leads to the well known algorithm of Giffler and Thompson (1963).

A1 Determine an operation o' from B with the earliest possible completion time, i.e. $t' + p' \leq t + p$ for all $o \in B$.

A2 Determine the machine m' of o' and build the set C from all operations in B which are processed on machine m'.

A3 Delete operations in C which do not start before the completion of operation o', i.e. $C := \{o \in C \mid t < c'\}$.

A4 Select that operation o^* from C which occurs leftmost in the permutation π and delete it from B, i.e. $B := B \backslash \{o^*\}$.

This algorithm produces schedules, in which no operation could be started earlier without delaying some other operation or neglecting any machine routing. Notice that active schedules are also semi-active.

Non-delay schedules are produced similarly by using a more rigid criterion for picking an operation from the critical set C. Modify step 2 as follows.

N1 Determine an operation o' from B with the earliest possible starting time, i.e. $t' \leq t$ for all $o \in B$.

N2 Determine the machine m' of o' and build the set C from all operations in B which are processed on machine m'.

N3 Delete operations in C starting later than o', i.e. $C := \{o \in C \mid t = t'\}$.

N4 Select that operation o^* from C which occurs leftmost in the permutation π and delete it from B, $B := B \backslash \{o^*\}$.

Non-delay scheduling means that no machine is kept idle when it could start processing some operation. Non-delay schedules are necessarily active schedules and hence also necessarily semi-active ones.

Non-delay schedules have a shorter makespan on average than active schedules. Concerning optimality, however, it is merely known that at least one schedule with minimal makespan is among the actives ones. As indicated in Fig. 7.9, there is in difference not necessarily an optimal schedule in the set of non-delay schedules.

For the above reason most adaptive approaches to the JSP engage an active schedule decoding, e.g. Yamada and Nakano (1992); Bierwirth (1995) and Dorndorf and Pesch (1995). Contrarily, a non-delay scheduler is used for decoding in the GA approach of Della Croce et al. (1995). They report that non-delay scheduling improves the solution quality for some problem instances whereas for other instances active scheduling succeeds. As a compromise the non-delay scheduler is augmented with a "look ahead" that widens the space of non-delay schedules. Without codifying all the active schedules the "look-ahead scheduler" accepts a machine running idle although it could start processing an operation in some cases.

The performance of EAs in job shop scheduling, which is reported in literature, is variable. Compared with neighborhood search, however, all approaches are outperformed in runtime as well as in solution quality.

2.3 GENETIC LOCAL-SEARCH

Due to this observation several hybrid variants of adaptive search have been proposed which incorporate features of neighborhood search. Pesch (1994) takes up ideas to solve the JSP through a decomposition by focusing on subproblems considering two jobs or single machines. In these approaches a GA is used to orient the disjunctive arcs between operations in the resulting graph formulations.

Recombination incorporating neighborhood search has been suggested by Aarts et al. (1994). In their approach all operations are identified in two parent schedules that might be swapped with respect to the critical-end neighborhood. Those pairs of operations existing in the second parent in inverse direction are reversed in the first parent. The new schedules finally undergo iterative improvements based on the critical-transpose neighborhood.

Another kind of genetic local-search is used by Mattfeld (1996). His algorithm engages permutations with repetitions of jobs for representing a problem instance. New candidate schedules are generated by an OX-like operator which preserves the particular repetition structure. These schedules are then subject to iterative improvements using the neighborhood definitions of Dell' Amico and Trubian (1993).

The results obtained by genetic local-search are clearly superior to those obtained by genuine adaptive search. The most effective appears to be the approach of Mattfeld (1996). Although the average quality of schedules is always comparing favorably with standard point-based search methods like Simulated Annealing and Tabu Search, the population-based hybrids are clearly outperformed in concern of the runtime needed.

2.4 COMPUTATIONAL STUDY

Throughout this chapter we have presented different ways of encoding and decoding solutions to scheduling problems. Now we may ask which way works best. Regarding the encoding issue, the priority representation of operations seems to capture the combinatorial nature of scheduling sufficiently well. What remains to decide on is the schedule-builder used.

Any decoding procedure basically determines the space that has to be searched. An unbiased decoding, which addresses the space of semi-active schedules, solely relies on the force of adaptation. Domain knowledge is incorporated by searching the space of active solutions. Alternatively, we can search the space of local-optimal schedules by employing an iterative improvement algorithm after semi-active decoding. These alternatives are investigated in this section by a computational study on the benchmarks described in Sect. 1.4. For this purpose we refer to the simple GA template sketched in Fig. 3.1 in the following.

1. The population size is set to 250 individuals.
2. Scaled fitness-proportionate selection is used.
3. The recombination step calls PPX crossover at probability of 0.6 in combination with pairwise-exchange mutations at probability of 0.1.
4. The algorithm is run for 100 iterations.

This setting is chosen with respect to the perfect preservation of precedence relations among operations caused by the PPX operator. As a consequence, an algorithm is able to exploit its schedule information very fast. To counterbalance the danger of premature convergence a comparably large population and a high mutation rate are used.

For a genetic local-search procedure we refer to the advanced order-based GA developed by Mattfeld (1996). Instead of the priority representation of operations, this algorithm uses the closely related representation of permutations with repetition of jobs. The schedule decoding is performed by an iterative improvement algorithm using a highly efficient neighborhood. Further parameters are set as follows.

1. The population size is set to 100 individuals.
2. Rank-based selection is used in a structured 10×10 population.
3. The recombination is controlled locally by an auto-adaptive strategy using order-based crossover (for details see Chap. 4(1.3)).
4. The algorithm is run for 100 iterations.

Table 7.6. Comparison of adaptive search algorithms using different decoding procedures (semi-active vs. active schedule-building vs. genetic local-search). Shown is the mean relative error against the best-known solutions in two problem classes.

problems	ϵ_{semi}	ϵ_{active}	ϵ_{gls}
"easy"	0.042	0.001	0.000
"hard"	0.515	0.221	0.226

Notice that the problem of premature convergence is circumvented in this implementation by means of the local recombination strategy involved.

For a comparison of the three adaptive search algorithms they are run 50 times each on the problems of our benchmark suite. The mean relative error to the best-known solutions are recorded and averaged over the five easy and hard to solve test problems respectively. Tab. 7.6 shows the results obtained.

Regarding the easy problems, an average relative error of approximately 4% is observed for the semi-active scheduling algorithm. Considering the hard problems it turns out that the algorithm badly fails. Recall from Tab. 7.2 that random schedules show an error of $\epsilon_{rand}=1.15$, i.e. adaptation does hardly halve this value. This means that adaptation is unable to navigate the search properly in case of a complex solution space.

The active scheduling algorithm and the genetic local-search algorithm are obviously in the same range of performance. The easy problems are solved to optimality by both approaches while an error of approximately 22% is measured for the hard problems. Applying an iterative improvement procedure after semi-active decoding reduces the relative error of semi-active scheduling by more than one half. Nevertheless, the genetic local-search algorithm is not superior to the active scheduling algorithm, although it takes advantage from several advanced EA techniques (e.g. structured populations). This observation is explained by our configuration space analysis for the JSP. We have found that fitness landscapes of the JSP show a plane overall surface where local optima of similar quality can reside almost everywhere. Thus, searching the space of local optima cannot be more effective than searching any other reduced solution space of comparable mean fitness. Presupposed a suitable encoding of schedules like the priority representation is used, the set of active schedules certainly addresses such a space. Therefore we conjecture that it is impossible to combine adaptive search and neighborhood search in a way, both can profit from each other.

Considering the computation costs, the building of active schedules is usually less expensive than obtaining local-optimality from semi-active schedules. This transfer corresponds to an adaptive walk in the fitness landscape. For the hard problems we have measured the average length of such walks, starting from randomly generated schedules, with 3,860 basic moves, see Tab. 7.4. While an EA progresses the mean fitness of the population increases and the effort spent for generating local-optimal schedules decreases at the same time. In contrast, an active schedule is obtained at constant costs. This makes it difficult to compare the runtime demand of the tested algorithms.

The average runtime of the active scheduling algorithm is about one minute for the hard problems. The corresponding runtime reported by Mattfeld (1996) for the genetic local-search algorithm is about ten minutes where only half the number of schedules are evaluated. Although measured on different computer systems, the comparison indicates that a simple GA can be at least as efficient for scheduling than a hybrid adaptive procedure which incorporates an iterative improvement algorithm.

2.5 DISCUSSION

Keeping in mind that the best-known solutions of our benchmarks have been generated by neighborhood-search algorithms, we must admit that adaptive search is clearly dominated for the hard JSP instances. This assessment is confirmed by two comparative studies on local search algorithms, performed by Vaessens (1995) and Anderson et al. (1997). In these studies a number of different approaches, including many of those described in Sect. 2.1 and 2.2, are applied to a large testbed. Although optimal solutions are known for many of the contained problems, all of them are extremely difficult to be solved to optimality. The average relative error of algorithms participating in the competition is determined against the best-known solutions respectively. The runtime of the algorithms, reported in literature, is standardized such that it is independent of the particular computer used.

Vaessens (1995) has plotted the observed performance against the runtime spent. The latter is given in logarithmic scale in Fig. 7.10. For clarity we group the points which result for the considered algorithms into four cluster, indicating the class of strategy they belong to.

It can be recognized that the performance of adaptive search (GAs) is only within the range of standard heuristics for the JSP, like the *shifting-bottleneck procedure* of Adams et al. (1988). The runtime demand of adaptive search is yet considerably larger. Even genetic local-search approaches hardly reach the solution quality achieved by genuine neighbor-

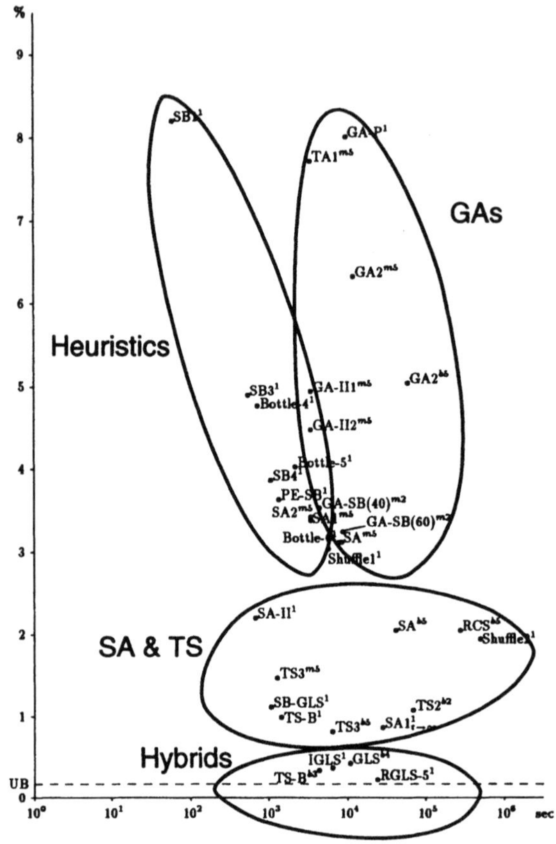

Figure 7.10. **Average relative error of local-search algorithms for another JSP benchmark suite versus the standardized computation time spent. The dashed line (UB) indicates the level of the currently best-known solutions.**

hood search techniques (SA & TS). Among the algorithms in this cluster Tabu Search is little superior to Simulated Annealing. Some of these algorithms are time consuming as well but they produce much better results than all adaptive scheduling approaches. Also neighborhood search appears to be hindered by the plane overall surface of JSP landscapes. This may explain the success of hybrid neighborhood-search algorithms which integrate limited Tabu walks into a backtracking framework.

3. SUMMARY

Recent investigations on local-search algorithms have revealed many innovative ideas for solving scheduling problems. The work already done has unquestionably evoked an enormous interest and progress. This is probably best observed at the example of the classical JSP. Nevertheless, research in the field of scheduling, as reported in the literature, predominantly concentrates on a few standard models including the JSP. These models contain by no means all of the complicating features faced in practice.

The transfer of local-search methods for real-world applications first of all provokes their flexibility. From this point of view EAs promise to handle many practical aspects more comfortable than neighborhood-search algorithms can. Adaptive search does not require to consider the relevant aspects of a problem in terms of a neighborhood definition. However, if an efficient neighborhood can be formulated, neighborhood search should be favored because EAs actually turned out to be rather weak methods in the field of scheduling. Combining ideas of both, adaptive search and neighborhood search unfortunately does not succeed.

Chapter 8

TOWARDS REAL WORLD SCHEDULING SYSTEMS

It has been shown within the last two decades that local-search based heuristics can fit the needs arising in many fields of combinatorial optimization. Mainly attracted by the difficulty of *NP*-hardness, academic research has concentrated on very simple models. Unfortunately, these model often do not contain the essentials of problems met in the real world. In the field of scheduling, an intensive discussion comprised the design of problem representations and search operators. But the suitability of the proposed techniques has been verified solely with respect to makespan minimization problems. Although other criteria of schedules are of more practical need, there is only little research on local search oriented towards pursuing these objectives. The transfer of experiences made for the JSP therefore remains an open challenge.

In this chapter we investigate the capabilities of the previously described search framework beyond the restrictive assumptions of the JSP. First, a number of additional elements of realistic scheduling scenarios is considered. In order to cope with an increasing complexity, a highly efficient schedule decoder is incorporated in the adaptive search. In the second section this new technique is applied to the problem of reschedule the planned activities in a changing environment.

1. SCHEDULING SCENARIOS

The section addresses the determination of schedules with regard to practical demands such as time-windows, service levels, and preferential treatment of important customers. Unfortunately, even these basic demands often prevent building an appropriate optimization model which allows the definition of efficient neighborhoods. Since neighborhood search is disabled, adaptive search becomes an attractive alternative for schedule optimization again. However, the capabilities of EAs to conduct a proper search still vanish with an increasing problem complexity. As a remedy an efficient schedule decoding procedure is proposed in the following.

1.1 LIMITATIONS OF THE DISJUNCTIVE GRAPH MODEL

Practical scheduling problems typically tend to be of dynamic rather than of static nature. In industrial production jobs are released for being processed with respect to capacity constraints, material supplies and costumer demands. Usually a *ready-time* is announced for a job which denotes the earliest possible starting time of processing the first operation of that job. A *due-date* is typically assigned to the jobs which represents a point in time at which a job should be completed in order to keep the projected date of delivery. The planning further considers *priorities* among the jobs already released for processing. Such priorities indicate the effort which has to be spent on completing a job in time with respect to the importance of a certain customer. We concentrate on these practical aspects by extending the basic JSP model introduced in Chap. 7(1.1) towards scheduling scenarios with

- non zero ready-times r_i of jobs,
- prescribed job due-dates d_i, and
- priorities among the jobs, expressed by weights $\omega_i > 0$.

In this context, makespan minimization is no adequate measure of schedule performance anymore, because the makespan crucially depends on the job with the latest ready-time. A flow-time dependent criterion can be used instead for schedule evaluation. This makes sense if a moderate work-in-process is pursued for reasons of flexibility or inventory-holding costs. The change towards customer-driven markets has led to situations where emphasize is often rather on due-date oriented criteria. This stems from the desire to improve costumer service or to avoid penalties for late deliveries.

Among a variety of criteria for schedule optimization, the following objectives are considered in this study

- the minimization of the weighted mean flow-time of jobs,
- the minimization of the weighted mean tardiness of jobs,
- the minimization of the maximum tardiness of jobs, and
- the minimization of the weighted number of tardy jobs.

A definition of the weighted mean flow-time refers to the completion time c_i of the jobs, which is determined by the starting time of the last operation plus its dedicated processing time. This figure is used in the well-known *weighted mean flow-time* measure

$$\overline{F} = \frac{1}{n}\sum_{i=1}^{n} \omega_i(c_i - r_i). \tag{8.1}$$

The tardiness of a job indicates the time span the completion time overshoots the projected due-date. Let $T_i = \max(c_i - d_i, 0)$ denote the tardiness of a job. Based on this measure, the *weighted mean tardiness* is defined as

$$\overline{T} = \frac{1}{n}\sum_{i=1}^{n} \omega_i T_i. \tag{8.2}$$

A related criteria, called the *maximum tardiness of jobs*, measures the worst violation of the due-dates

$$T_{\max} = \max\{T_i \mid 1 \leq i \leq n\}. \tag{8.3}$$

Finally, let $tardy(i)$ indicate whether job i is tardy or not, i.e. $tardy(i) = 1$ means it is tardy ($T_i > 0$) while $tardy(i) = 0$ means it is completed in time. Then, the *weighted number of tardy jobs* is simply given by

$$T_{\mathrm{n}} = \sum_{i=1}^{n} \omega_i tardy(i). \tag{8.4}$$

As it can be recorded easily, this measure is frequently met in practice. It is of particular interest also, because it has a direct influence on the logistical *β service-level* which designates the percentage of on-time deliveries.

Let us now concentrate on changes to the optimization model and the impact on local-search methods. It has been shown by White and Rogers (1990) that additional data of ready-times, due-dates and job-weights can be easily included in the graph formulation of the basic JSP.

This may happen by e.g. assigning the ready-time of a job to its initial arc starting from the begin node, compare Chap. 7(1.2). Similarly, the representation of due-dates and weights introduces further nodes to the corresponding graph. In this way the graph formulation is extended towards additional data.

Considering different criteria than the makespan, it is obvious that the critical path cannot serve as a measure of performance anymore E.g. for $\overline{F}$ and $\overline{T}$ multiple critical paths must be considered for each job independently[1].

The central idea of neighborhood search, which is pretty successful for the basic JSP, is to consider a subset of the possible perturbations of a schedule by taking the critical path into account. This strategy enables the definition of small neighborhoods which yield improvements with high probability while ensuring feasibility at the same time. If the objective of a scheduling problem is not a function of the critical path, however, the reversal of a non-critical disjunctive arc can possibly improve a schedule as well. Of course, the feasibility claim does by no means hold for such kind of perturbations. Moreover, lower-bound estimations of the objective-function value, which can bypass building the new schedule from the scratch, are not available anymore. In consequence, neighborhood search requires evaluating extremely large neighborhoods at enormous computational costs.

The lack of efficient neighborhood definitions for realistic scheduling problems has renewed the research interest in EAs. Obviously, while making a single step in a neighborhood search, adaptive search can progress numerous iterations. This argument is hoped to remedy the apparent weakness of EAs in scheduling.

1.2 BENCHMARK PROBLEMS

For a computational study we concentrate on a benchmark suite provided by Morton and Pentico (1993). This testbed extends the basic JSP benchmark problems considered so far towards scenarios which emphasize real-world demands. The entire suite consists of twelve problems of which only six are treated here for reasons of comparison with other EA approaches.

It can be taken from Tab. 8.1 that the problems no. 1 and 2 consist of thirty jobs each, to be scheduled on three machines. The larger problems no. 7, 9, 10 and 12 contain about three times the operations of the smaller problems. The total number of operations is slightly smaller than $n \times m$

[1] cf. Rixen (1997).

Table 8.1. A heterogeneous testbed of scheduling benchmarks which incorporates diverse features of realistic problems.

problem no.	1	2	7	9	10	12
number of jobs n	30	30	50	50	50	50
number of machines m	3	3	5	5	8	8
operations in total	67	60	193	211	250	230
average allowance A	1.4	1.5	1.5	1.5	1.4	1.5
utilization rate U	0.7	0.9	0.8	1.0	0.7	0.9
std. dev. of machine load	0.1	0.4	0.2	0.3	0.1	0.3
std. dev. of job weights	0.5	0.6	0.3	0.5	0.6	0.4

which indicates that the jobs do not necessarily pass every machine. Different than the problems considered in the benchmark study of the previous chapter, these scenarios determine dynamic problems because all jobs carry ready-times and due-dates.

The time span between the ready-time and the due-date of a job is called its *allowance*. Let $P_j = \sum_i p_{ij}$ denote the total processing time of a job. The allowance can be given in percent of the total processing time

$$A_j = \frac{d_j - r_j}{P_j}, \tag{8.5}$$

expressing the tightness of a job's due-date. Obviously, if $A_j < 1$ a job will always be tardy. However, completing a job in time if A_j is hardly larger than 1 is not very likely as well. An allowance of about 1.5 of the total job processing time appears to indicates rather tight due-dates.

Considering due-dates alone does not sufficiently reflect the complexity of dynamic scheduling problems. Also the setting of ready-times decides on the difficulty of scheduling. Rapid job releases lead to a high workload, because many jobs concurrently await processing at a certain point in time. Jobs that await processing by the same machine are conflicting. Obviously, the more such conflicts arise, the more difficult a problem gets.

The observed workload is caused by two independent forces, the arrival process of jobs and the scheduling process itself. The former force is described by the *mean inter-arrival-time* of jobs

$$\lambda = \frac{1}{n-1} \sum_{j=1}^{n-1} (r_{j+1} - r_j). \tag{8.6}$$

Due to the unknown quality of scheduling, the workload cannot be prescribed for a problem. Therefore the *machine utilization rate* is used to approximate a certain workload situation by neglecting the influence of scheduling. Let $\overline{P}$ denote the average processing time of jobs, i.e. $\overline{P}=\sum_j P_j/n$. Since at most m machines are busy in parallel, the utilization rate is given by

$$U = \frac{\overline{P}}{m\lambda}. \tag{8.7}$$

A value $U > 1$ leads to a continuous increase of the workload over time. With $U = 1$ the jobs are released so rapid that already the operation processing times fully occupy the machines. An increase of workload can be prevented only, if (i) machine idle-times can be avoided theoretically and (ii) the scheduling process is able to exhaust the machine capacities totally. For a utilization ranging from 0.7 to 1.0, challenging scenarios can be expected in general.

Another typical observance of scheduling scenarios is introduced into the testbed by the distribution of operation processing times. They do not occupy the machine capacities consistently as in most of the standard JSP benchmark-suites. This is shown by the load deviation of the machines. A heavy loaded machine will probably turn out as a bottleneck during the scheduling process. The last line of the table refers to the job-weights which are uniformity distributed to different means in the problems respectively. All this verifies a strong heterogeneity of the testbed.

1.3 PROBABILISTIC SCHEDULING

Practical scheduling methods predominantly base on *priority rules* which are found to work favorable by experience[2]. Roughly spoken, a priority rule is a strategy that selects a job among those awaiting processing each time a machine gets idle. For a simulation of this process we can apply a non-delay schedule-builder where selecting from conflicting operations is based on a particular priority rule.

Among the large number of rules proposed in the literature we concentrate on three rules which are known to reduce job flow-times and job tardiness effectively. The FCFS rule is considered for reasons of comparability.

FCFS. The *first-come, first-serve* rule attempts to implement a fair conflict solver. The outcome of this rule is close to what can be expected from reasonable decisions while neglecting any particular properties of jobs.

[2]These components of shop-floor control have been intensively studied over the last decades, see e.g. Holthaus (1996) for a recent survey.

Table 8.2. Average performance of probabilistic priority-rules on the benchmarks regarding different objectives. Shown are the relative improvements gained against deterministic FCFS scheduling.

rule	$\overline{F}$	$\overline{T}$	T_n	T_{max}
FCFS	0.000	0.000	0.000	0.000
$FCFS_p$	0.046	0.099	0.097	0.135
S/OPN_p	0.023	0.170	0.013	0.327
SPT_p	0.083	0.198	0.205	0.075
$COVERT_p$	0.051	0.217	0.129	0.269

SPT. The *shortest processing time* gives priority to that operation with the shortest imminent processing time. Jobs waiting in a queue may cause that their successor machines run idle. SPT alleviates this risk by reducing the length of the queue in the fastest possible way.

S/OPN. The *slack* of a job is defined as the time span left within its allowance, assuming that the remaining operations are performed without any delay. Since jobs may wait in front of each machine the rule *slack per number of operations remaining* gives priority to the job with the minimum ratio of slack and the number of remaining operations.

COVERT. The *cost-over-time* rule combines ideas of SPT and S/OPN It prioritizes jobs according to the largest ratio of the expected job tardiness and the required operation processing-time. In this way it retains SPT performance but seeks to respect the due dates if jobs are late.

Simple modifications of the above rules enable considering the job-weights explicitly. Notice that a deterministic schedule-builder produces a single schedule on the basis of a priority rule. According to Baker (1974), a more balanced assessment of a rule results from applying it in probabilistic fashion. For this end the priorities assigned to conflicting operations are used for stochastic sampling. In this way, multiple schedules are produced from one rule by retaining its original character.

To get insight into the intractability of the test suite the above rules are probabilistically applied to each of the problems in 1,000 iterations. In order to report aggregate measures for the entire benchmark suite, the best result gained for a certain problem is set in relation to the outcome of the deterministic version of FCFS. These values, averaged over the six problems, are shown in Tab. 8.2 with respect to the four criteria under consideration.

The results confirm that a probabilistic scheduler can produce considerable quality improvements. For the testbed they range from approximately 8% to 32%. SPT is well-known to be the dominating rule if flow-time minimization is pursued. This rule also works quite well in reducing the weighted number of tardy jobs. Nevertheless, for other due-date based criteria SPT performs rather weak because it tends to delay those jobs badly which require comparably large processing times on at least one machine. S/OPN shows to be favorable for minimizing the maximum tardiness of jobs. Here, SPT fails completely and is even outperformed by probabilistic FCFS. In order to minimize the mean tardiness of jobs the COVERT rule is most suitable. In summary we can state that no general purpose rule is available which motivates further research concerning robust scheduling techniques.

1.4 ROBUST ADAPTIVE SCHEDULING

The GA which has been developed for basic scheduling problems can be directly applied to the new testbed by (i) switching the fitness function with respect to the desired measure of performance and (ii) taking care of ready-time while building candidate schedules. This is possible because all of $\overline{F}$, $\overline{T}$, T_n, and $T_{\max}$ belong to the important class of *regular* performance measures[3]. In order to minimize a regular measure it is necessary to consider active schedules only[4]. In other words, all problem instances have at least one optimal schedule where no operation could start earlier without delaying any other operation or violating the prescribed machine routings. Adaptive search can therefore take advantage from the schedule decoding procedures introduced in Chap. 7(2.2).

To simplify matters, we further rely on a similar parameter setting of the GA as in the previous computational study.

1. The population size is set to 100 individuals.

2. Scaled fitness-proportionate selection is used.

3. The recombination step calls PPX crossover at probability of 0.6 in combination with pairwise-exchange mutations at probability of 0.1.

4. The algorithm terminates after τ iterations are carried without gaining any improvement. The parameter τ is set to half the number of operations contained in the particular scheduling problem.

---

[3] A criterion is referred to as regular if the objective function value is non-decreasing in the job completion times. Suppose two schedules where the jobs complete no later in the first schedule than in the second schedule. Then, under a regular measure of performance, the first schedule is as least as good as the second.

[4] For a proof of this theorem see e.g. French (1982).

Table 8.3. Comparison of performance between probabilistic rule-based scheduling and two variants of adaptive scheduling.

schedule quality	$\overline{F}$	$\overline{T}$	T_n	T_{max}
best probabilistic rule	0.083	0.217	0.205	0.327
mean GA (non-delay)	0.095	0.279	0.259	0.314
mean GA (active)	0.089	0.294	0.295	0.322
standard deviation				
best probabilistic rule	0.02	0.07	0.06	0.11
mean GA (non-delay)	0.01	0.02	0.02	0.03
mean GA (active)	0.02	0.04	0.04	0.05
average runtime (sec.)				
GA (non-delay)	14	15	7	9
GA (active)	30	30	9	14

Since the problems of the new testbed are considerably smaller than the JSP benchmarks investigated throughout the previous chapter (at least for the number of operations in total), the population size is reduced from 250 to 100 individuals. As a further difference, a flexible termination criterion is used with respect to the heterogeneity of the testbed. In this way more complex problems is given reasonable longer time to converge.

The above setting is tested in two variants which incorporate different schedule decoding techniques. Recall that priority rules have been evaluated on a non-delay scheduling basis. To achieve easy comparability, a non-delay scheduler is also applied by one GA variant. The other GA variant performs active schedule building. Both algorithms are run for 50 times on the six benchmark instances and the four different objectives each. The results gained are aggregated like described above. Tab. 8.3 shows the outcome and runtime[5] of these computations together with the result of the best performing rules.

Both GA approaches adapt towards the requirements defined by the different performance criteria. With the only exception of one criterion, they even outperform the most suitable priority rule available. The improvements range from 1% for $\overline{F}$, over 7.5% for $\overline{T}$, to 9% for T_n. This underlines that SPT is pretty effective for reducing flow-times. Regarding T_{max}, however, neither the non-delay GA nor the active GA is able to produce a schedule as good as S/OPN. A striking observation is that

[5] measured of a Pentium/200 Mhz computer.

the active GA is superior to its non-delay counterpart merely for three criteria. For $\overline{F}$ the non-delay GA turns out clearly advantageous.

The standard deviation of the GA results is pleasantly small if compared with probabilistic scheduling. A small deviation is considered advantageous because an adaptive search method, different than a priority rule, might be applied only once under realistic conditions. We reflect this handicap by reporting the mean schedule quality achieved. The observed standard deviation consistently duplicates when switching from non-delay to active scheduling. Due to the flexible termination criterion used, it is noticed that the active GA is more time consuming on average than its non-delay counterpart. This is explained by the larger search space, explored under an active schedule-builder. The increase of runtime is similar compared to the standard deviation observed for both adaptive approaches.

The computation time of the rule-based approach cannot be compared, because it depends on the number of iterations carried out by probabilistic scheduling. Nevertheless, it should be mentioned that the effort spent for the stochastic sampling process is not negligible.

The robustness of adaptive scheduling concerning different measures of performance appears highly attractive. Although quite successful if compared with well-known priority rules, it must be doubted whether the current state of algorithm can yield near-optimal schedule quality. In the previous chapter adaptive search is judged as an apparently weak strategy for scheduling. Thus far nothing has been added to cause substantial improvements.

1.5 IMPROVING ADAPTIVE SCHEDULING

The most common way to improve EA performance focuses on tuning the parameters of the algorithm, i.e. to enlarge the population size, to alter the selection pressure, etc. If all the parameters are already within useful bounds, the promised progress for solving a single problem instance usually does not justify the costs of finding an even more appropriate fine tuning. More universal approaches are going by modifications of the population management. Although coming along with slight improvements, such techniques are certainly not able to eliminate the weakness of adaptation concerning a particular problem class. Any substantial improvement of adaptation basically requires a modification of the scope of search. This may happen either by modifying the problem representation or by intensifying the decoding procedure.

Following the latter direction, Della Croce et al. (1995) have noticed that a non-delay scheduler sometimes produces much better results than an active one. In the above study we have made a similar observation for

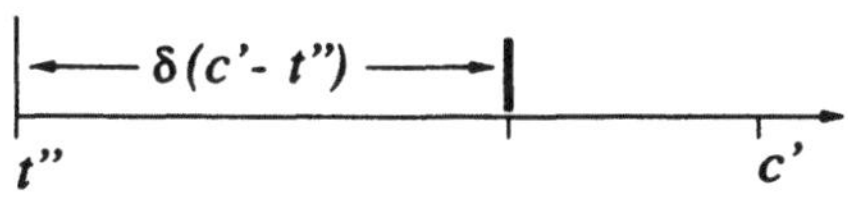

Figure 8.1 Variable schedule building on machine m' with $\delta = 2/3$.

$\overline{F}$. This finding strikes because it is known that non-delay schedules form a true subset of the active schedules. Conversely, there may exist active schedules of better quality than the best non-delay schedules. Although within the scope of search, solutions offered by an active scheduling GA are sometimes even worse than those found by a non-delay scheduling GA. In these cases adaptation obviously fails to explore a larger search space.

This discovery motivates the idea to vary the scope of search performed by the scheduler. The variable building of schedules is enabled by introducing a tunable parameter $\delta \in [0,1]$. The setting of δ can be thought of as defining a bound on the time span a machine is allowed to remain idle. At the extremes $\delta = 0$ produces non-delay schedules, while $\delta=1$ produces active schedules. A simple version of this schedule-builder[6] is given in terms of the framework introduced in Chap. 7(2.2) for decoding permutation representations of schedules.

Variable schedules are produced by considering the set B, containing of all unscheduled operations whose job-predecessors are already scheduled.

V1 Determine an operation o' from B with the earliest possible completion time $c' = t' + p'$, i.e. $c' \leq t + p$ for all $o \in B$.

V2 Determine the machine m' of o' and build the set C from all operations in B which are processed on machine m'.

V3 Determine an operation o'' from C with the earliest possible starting time, i.e. $t'' \leq t$ for all $o \in C$.

V4 Delete operations in C which do not start before the maximal span of idle time $\delta(c' - t'')$ of machine m', i.e. $C := \{o \in C \mid t'' \leq t < t'' - \delta(c' - t'')\}$.

V5 Select that operation o^* from C for being scheduled next which occurs leftmost in the permutation π and delete it from B, i.e. $B := B \backslash \{o^*\}$.

This tunable decoding procedure generates schedules based on a variable *look-ahead.* All operations which could start within the maximal

[6] For details of an efficient implementation of this algorithm in order $O(n^2)$ see Storer et al. (1992).

Table 8.4. Mean performance of adaptive search using a variable schedule-builder.

δ	$\overline{F}$	$\overline{T}$	T_n	T_{max}
0.0	0.095	0.279	0.259	0.314
0.2	0.104	0.319	0.276	0.345
0.4	0.111	0.335	0.291	0.356
0.6	0.113	0.349	0.314	0.372
0.8	0.106	0.334	0.308	0.353
1.0	0.089	0.294	0.295	0.322

allowed idle time of a machine (shown in Fig. 8.1) have equal chance for being performed next by that machine. By varying δ within its bounds, the scope of adaptive search is continuously scaled from non-delay schedules to active schedules, compare Fig. 7.9.

In order to evaluate this approach we concentrate on the testbed again. Using the same experimental setup described above, now a variable scheduler is employed by the adaptive search algorithm. Its δ parameter, ranging in the interval $[0, 1]$, is increased by steps of 0.2 for this purpose. The results are shown in Tab. 8.4. Here, the first and the last line (non-delay schedule building with $\delta = 0.0$ and active schedule building with $\delta = 1.0$) are taken from Tab. 8.3.

The most striking result of this study is that average solutions, better than those obtained from non-delay and active scheduling, are found in every case. Regardless of the particular objective pursued, a value of δ=0.6 yields the best average performance for the entire testbed. Notice that even the notorious T_{max} criterion, where S/OPN has been the best method thus far (T_{max}=3.27), is improved clearly by 4.5% under δ=0.6. SPT, so powerful concerning the $\overline{F}$ criterion, is outperformed by at least 3%. For the remaining criteria $\overline{T}$ and T_n, improvements larger than 10% are achieved against the best performing rules respectively.

With respect to the heterogeneity of the testbed, it can be conjectured that for some problem instances, values different than $\delta = 0.6$ work best. To investigate this question we take a look at problems no. 7, 10 and 12 while varying δ. Thereby it is exemplary focussed on the $\overline{T}$ criteria. In order to achieve comparability for the three problem instances under consideration, the measure is normalized in the following way. The worst average performance, produced by 50 GA runs under a particular value of δ for the problem, is assigned a normalized measure of 0.0. Accordingly, the best average performance is assigned a normalized mean tardiness of 1.0. Further results which are obtained for a problem using

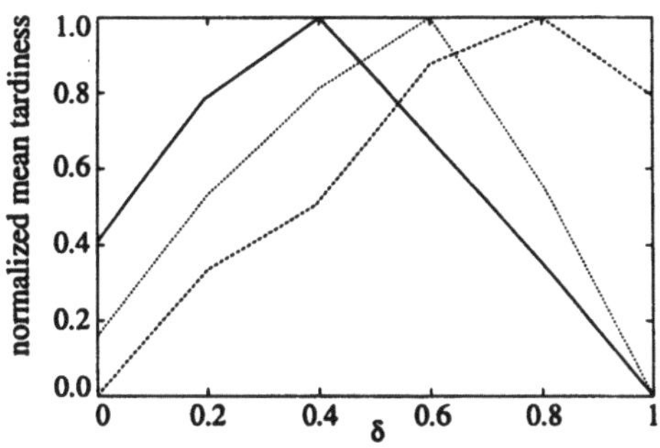

Figure 8.2
GA performance, depending on δ, for three selected benchmark problems.

a certain value of δ are linearly scaled in the range of 0.0 and 1.0. This leads to the plot shown in Fig. 8.2.

The diagram confirms the conjecture made above that the proper functioning of GAs for scheduling is limited to a certain size of the search space. Starting from non-delay scheduling with $\delta = 0$, the increase of δ leads to improved solutions for all three problems. Previously excluded solutions are now included into the scope of the search while the GA still works sufficiently well. Beyond a certain value of δ, a further increase leads to continuous deterioration of the solution quality, although potentially better solutions are still included into the search space. In other words, a trade-off is observed between the incorporation of a few solutions of superior quality and the vast majority of inferior solutions.

We find different values of δ to be most suitable for the three problem instances. These critical values, where the potentials of the GA are exceeded, indicate the specific intractability of the instances. The smaller such a value is, the more difficult we expect the instance to be solved.

Finally, it is worthwhile to mention that the runtime demand of adaptive search strongly depends on the look-ahead taken while building a schedule. Using an active scheduler ($\delta = 1$) requires the largest runtime and using a non-delay scheduler ($\delta = 0$) requires the shortest. The corresponding runtime can be taken from Tab. 8.3. Within these bounds, the measured runtime increases approximately linear with the parameter δ.

To summarize, we have seen that active scheduling does not always produce the best results. The reliability of adaptation obviously depends on the complexity of the space to be searched. Fortunately, the scope of this space can be scaled by decoding of schedules. This enables to find solution of satisfying quality at reasonable computational costs even for very complex problems.

2. RESCHEDULING

Detailed schedules can usually not be legislated in advance, they rather emerge from a system's incessantly changing environment. For instance in car manufacturing, Van Dyke Parunak and Fulkerson (1994) estimate the practicability of a daily schedule by an hour or less. Within this time span so many unforeseen events have occurred that it is impossible to execute a schedule any further. In distribution systems which depend on road traffic, the lifespan of schedules will hardly last longer. Thus, accepting that schedules are rarely implemented like planned, the problem is less of finding an optimal operation sequence for later execution, than of coping with the dynamic unpredictability of the environment.

The most effective approach to practically handle non-deterministic environments is monitoring the mismatch between the predicted and the actual state of a system. Whenever the observed deviation exceeds some threshold, the current plan is *rescheduled* in order match the planning data with reality again. Rescheduling is possible only when the time needed to revise a plan is short compared with the time span between a change in the environment and the need to react on this change. Therefore runtime efficiency is a primary prerequisite of rescheduling algorithms.

2.1 CHANGES TO THE MODEL

Scheduling models typically assume that there is a fixed number of n jobs to be scheduled. After having scheduled these jobs, a problem is viewed as being solved. When considering a real-world problem, there may be n jobs in a system at any time, but further jobs are continuously released. Thus, scheduling the available jobs has to be done under uncertainty concerning events in the near future. The release of a new job can be such an event, but it can also be an unexpected machine-breakdown, a change of job preferences (a job of low priority may suddenly become high priority), and the like. In practice, the occurrence of unforeseen events require to modify the existing schedule. This process is referred to as *reactive scheduling* or simply as *rescheduling.*

In order to employ adaptive search for rescheduling, we concentrate on a simplified non-deterministic production environment. In this system new jobs are released from time to time. After each such event a new scheduling problem can be formulated with respect to all operations currently known in the system[7]. In this way a series of deterministic scheduling problems is produced to reflect the non-deterministic production environment.

[7]E.g. the test problems tackled in this chapter can be thought of as scheduling scenarios which result from the release of the job with the latest ready-time.

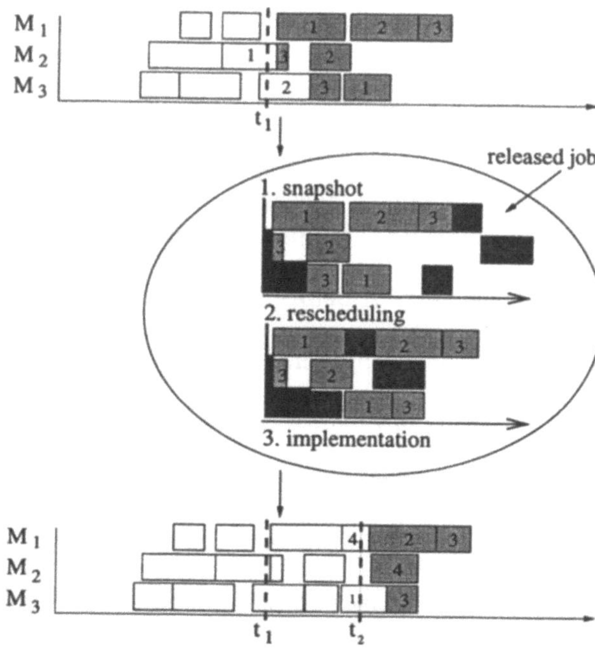

Figure 8.3. Production scheduling on a rolling time basis.

Let P_0 be a deterministic scheduling problem. Solving this problem leads to a table of potential starting times for all operations involved. At the release of a new job at time t_1, problem P_1 is derived from this current schedule. To construct the new scheduling problem, operations having potential starting times $t < t_1$ are thought of as being completed. Consequently, they are removed from the problem. If the last operation of a job has been removed this way, the corresponding job is removed completely. For jobs which have been modified but not removed, a new ready-time is determined by the completion time of their last operation removed.

Figure 8.3 shows a snapshot of a production system consisting of three machines at time t_1. Operations which have not been started before t_1 appear in light grey shade. Sometimes the processing of an operation, starting shortly before a new job has been released, overshoots t_1. The corresponding machine is busy at time t_1 and therefore not available. In the charts this is depicted by black shadings on M_2 and M_3. To model these busy periods of machines, *lead-times* s_j are introduced for the machines, denoting the earliest point in time machine j is available again. Similar to the ready-times of jobs, machine lead-times can be handled by the schedule-builder. The new job released at t_1 (shown in dark grey shade in Figure 8.3) is finally added to the remaining production program. This updated program finally results in the new scheduling problem which has to be solved next.

The particular steps of updating a scheduling problem in time are summarized below. This procedure changes a scheduling model dynamically with respect to the changes observed in the environment[8]. After each job release (say at time t) the following steps are carried out.

1. Remove all operations started before t from being considered.
2. Remove a job if all its operations have been removed already.
3. Adjust the ready-times r_i of jobs which are begun but not completed.
4. Adjust the lead-times s_j of machines which are not available at t.
5. Add jobs released at t to the problem data.

Obviously, this *snapshot procedure* can be applied at any time, independent of whether new jobs have been released or not. In case that no further jobs are added in step 5, the procedure simply causes a complexity reduction for scheduling of the remaining jobs. Due to this functionality the procedure can be applied periodically. By simple modifications, it can also be used in reaction to further typical events in production systems[9]. For simplicity, however, we only consider non-deterministic job releases in the following.

2.2 ADAPTIVE MEMORY ACCESS

Of course, any available algorithm which is capable to solve dynamic scheduling problems can be used for rescheduling as well. A simple way to employ such an algorithm is to monitor a production system as sketched in Fig. 8.3. After every system snapshot, the algorithm produces a schedule for the current state of the system. This schedule is implemented in the system as far as possible while the backlog is included in the reformulation of the scheduling problem. Then the algorithm is restarted again.

In real-world problems the used scheduling algorithm must work sufficiently fast to cope with the frequency of changes in the environment. In the following we therefore present an integrative approach to rescheduling which is based on an adaptive search component. The central idea of

[8] cf. Bierwirth and Mattfeld (1999) for a formal treatment of changing the model.

[9] Sudden machine breakdowns are treated e.g. in step 3 by temporarily setting the lead-time of the corresponding machine j to $s_j = \infty$, see Kopfer et al. (1995).

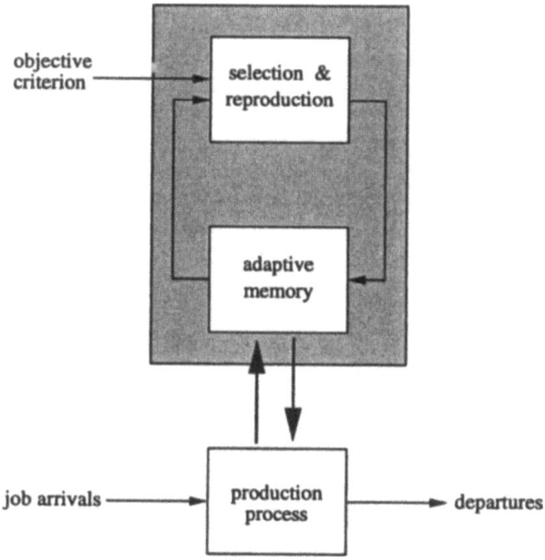

Figure 8.4 The GA structure for rescheduling with adaptive memory access.

this highly efficient rescheduling procedure was firstly mentioned by Fang et al. (1993). It is to reuse the already scheduled backlog of operations within the rescheduling process. After a snapshot is taken from the production system, a new problem is created like described above. Different to the previous approach, the population of the newly started GA is now obtained from the final population of the previous GA by adjusting it to the requirements of the new problem. In other words, the GA receives an image modification of its internal model, compare Chap. 2(1.3). In this way the search process is given a bias instead of initializing it at random. The population now acts like an *adaptive memory* in the rescheduling process.

This principle uses the basic communication structure of GAs as depicted in Fig. 5.1. It is adopted in Fig. 8.4, where a production process is viewed as an object system which is loosely coupled with an adaptive search. The search process is suspended after a snapshot is taken from the production system at a certain state. Afterwards the GA-population is adjusted according to the observed state of the system. The *population update* requires two simple steps.

1. Operations which have been started, including those which are currently processed, are deleted in all permutations of the GA population.

2. Operations of new jobs are randomly inserted in all permutations by leaving their order of precedence unchanged.

Table 8.5. Mean performance of adaptive rescheduling for a variable degree of problem decomposition. The schedule-builder is parameterized with $\delta = 0.6$.

ϵ	$\overline{F}$	$\overline{T}$	T_n	T_{max}
1	0.113	0.349	0.314	0.372
2	0.115	0.348	0.319	0.382
3	0.115	0.345	0.313	0.384
6	0.114	0.340	0.300	0.381

After these modifications, the GA continues the search process. The initial population now contains fragments of favorable schedules generated in the previous GA run. It can be conjectured that these fragments resemble each other strongly, since the previous population is likely to have nearly converged. Biasing the population will therefore lead to a much faster convergence which in turn results in a considerable saving of runtime. Of course, the more backlog we observe in the periods, the more advantageous adaptive memory will become.

2.3 ADAPTIVE RESCHEDULING

For a computational study on rescheduling we again focus on the testbed presented in Tab. 8.1. The existence of job ready-times allows a decomposition of the instances into a number of consecutive problems. First, all jobs involved in a problem are sorted by increasing ready-times. Next, this list is split into a prescribed number of approximately equally sized sublists which represent the planning periods. A planning period starts at the ready-time of the first job of its sublist of jobs. The period corresponding to the current problem is finally built from this sublist and the backlog of operations which have not been processed in the previous period.

The degree of decomposition which is used to solve a dynamic scheduling problem determines the size of the space that has to be searched in each of the periods. Let ϵ denote the degree of decomposition by the number of involved periods. Obviously, the larger ϵ is, the smaller the search space gets. We suppose that small problems of the periods can be solved to near-optimality. On the other hand, enlarging ϵ also reduces the horizon of planning which worsens the overall quality of scheduling. In order to discover a reasonable trade-off between the size of the search space and the benefits of planning, the decomposition degree is varied by $\epsilon = 1, 2, 3, 6$ in the following experiments.

Like in the previous studies, the GA is run for 50 times on the six test problems and the four different objectives each. When $\epsilon = 6$ it is

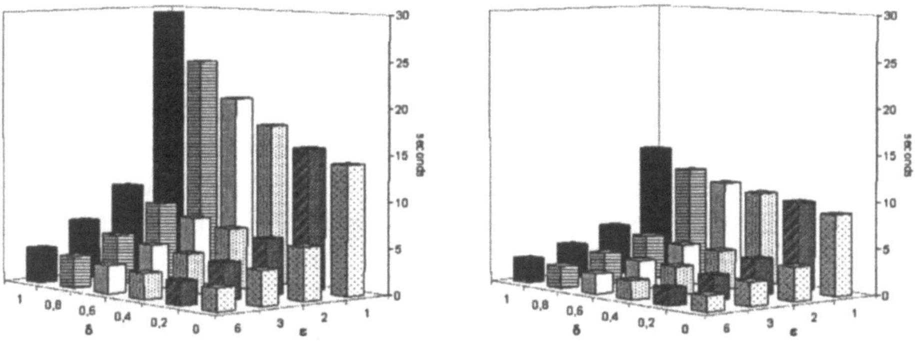

Figure 8.5. Average runtime (sec.) of adaptive rescheduling needed to solve a single test problem for $\overline{F}$, $\overline{T}$ (left) and for T_n, T_{max} (right).

measured that the test problems permit to carry over about 10% of the detailed schedules from one period into the other. For smaller values of ϵ this percentage certainly decreases. Since this small backlog can hardly result in a significant search bias, the GA is used without adaptive memory access. Like before, the look-ahead parameter of the schedule decoder is varied from non-delay scheduling ($\delta = 0$) to active scheduling ($\delta = 1$) by a step-size of 0.2. As in the previous computations a value of $\delta = 0.6$ turned out to work best on average again. The aggregated results obtained for this setting are shown in Tab. 8.5 with respect to a variable degree of problem decomposition.

Under $\epsilon = 1$ the test problems are solved at a whole. The corresponding results in the first line of Tab. 8.5 are therefore already known from the line shown for $\delta = 0.6$ in Tab. 8.4. It can be seen clearly that a further increase of ϵ hardly effects the solution quality gained. In some cases, a higher problem decomposition allows to increase the overall schedule quality slightly. Under $\epsilon = 6$, however, this tendency is saturated completely.

Let us now have a look at the runtime needed to solve the test problems under a variable degree of problem decomposition. As a result of Sect. 1.5, the time spent for schedule adaptation increases approximately linear with the parameter δ which is used for decoding schedules. This has been explained by the enlargement of the search space. Of course, increasing ϵ has an opposite effect as it actually reduces the scope of search by a problem decomposition. Thus one can expect savings of runtime by increasing ϵ.

The above conjecture is confirmed by Fig. 8.5. It shows the average GA runtime for the test instances regarding the possible combinations of δ and ϵ. The mean criteria $\overline{F}$, $\overline{T}$ and T_n, T_{max} are not distinguished because the GA requires an almost identical runtime for the objectives of each group. It is nearly twice as long for $\overline{F}$, $\overline{T}$ than for T_n, T_{max}. This figure indicates that the adaptation process generates twice the number of improvements for the mean performance criteria considered.

For a final consideration on the runtime demand we compare our approach with the traditional way of applying GAs to a scheduling problem. Here, candidate schedules are built on an active basis while the problem is solved at a whole. This means that δ and ϵ are set to a value of 1 each. Regarding the problems of our testbed this leads e.g. to a mean measure of 0.294 for $\overline{T}$ and requires approximately 30 seconds of runtime. Turning to a variable decoding in combination with a period decomposition, the solution quality is improved by about 18% while only 5 seconds are needed. The plots of Fig. 8.5 impressively depict the benefit gained. Already cutting the problems into two periods effects a shortening of runtime larger than is obtained by changing from an active schedule-builder to a non-delay schedule-builder.

2.4 COMPARISON WITH OTHER APPROACHES

In order to assess what has been gained so far we compare the performance of our GA with other evolutionary approaches to dynamic job shop scheduling. In two earlier papers published by Fang et al. (1996) and Lin et al. (1997) the same testbed has been tackled by a detailed consideration of the six test problems. Therefore we now provide the best results found in the previous computational study. Recall that 50 runs of our algorithm have been performed under all combinations of δ and ϵ for each instance. Actually, different values of δ and ϵ turned out to be favorable for the instances respectively, but they are not discussed here[10].

Fang et al. have applied two different adaptive approaches to the testbed. The first method is a stochastic hill-climbing algorithm which works quite similar to an ES with one parent and one offspring. This algorithm performs approximately 35,000 evaluations of schedules. As a second method, the GA previously described by Fang et al. (1993) is employed for schedule optimization. In order to provide a universally applicable algorithm, tailored operators have been omitted in this approach in favor of standard techniques such as a permutation representation and uniform crossover. For each of the test problems under investigation at most 100,000 active schedules are built. Thus we can

[10]For more details see Mattfeld (1999).

Table 8.6. Performance of adaptive scheduling approaches on the test problems of Tab. 8.1. The best found schedules, measured by the improvement-rate against deterministic FCFS scheduling, are compared for the various objectives. Superior measures are shown in grey shade for the problems respectively.

criterion	problem no.	best rule	FCR96	LGP97	MB98
	1	0.029	0.099	0.099	0.099
	2	0.156	0.182	0.182	0.182
$\overline{F}$	7	0.088	0.075	0.120	0.131
	9	0.179	0.166	0.229	0.254
	10	0.095	0.084	0.107	0.140
	12	0.117	0.109	0.145	0.182
	1	0.089	0.380	0.388	0.379
	2	0.442	0.475	0.491	0.484
$\overline{T}$	7	0.451	0.404	0.670	0.688
	9	0.334	0.272	0.486	0.499
	10	0.308	0.352	0.374	0.428
	12	0.393	0.361	0.473	0.568
	1	0.167	0.334	0.428	0.423
	2	0.451	0.550	0.551	0.545
T_n	7	0.334	0.429	0.627	0.618
	9	0.413	0.591	0.688	0.750
	10	0.154	0.300	0.319	0.315
	12	0.358	0.609	0.615	0.638
	1	0.383	0.596	0.585	0.596
	2	0.591	0.666	0.666	0.665
T_{max}	7	0.478	0.500	0.646	0.736
	9	0.383	0.455	0.574	0.543
	10	0.442	0.455	0.526	0.539
	12	0.660	0.656	0.685	0.713

state that the basic principles as well as the effort spent on building schedules are very similar to the algorithm used in the previous study. A fair comparison of performance should therefore be possible.

In contrast to Fang et al., Lin et al. have tackled the testbed by a highly sophisticated GA. This algorithm takes advantage from a whole bunch of evolutionary computation techniques, such as a direct problem representation which encodes the operation starting times, tailored genetic operators, and an island-structured population model. To cope with the high computational demand, this algorithm is parallelized. This feature allows Lin et al. to performs 1.125 million evaluations of schedules in each run.

The best solutions found by the approaches of Fang, Corne and Ross (FCR96), by the GA of Lin, Goodman and Punch (LGP97), and by the adaptive algorithm of Mattfeld and Bierwirth (MB98) are shown in Tab. 8.6. For a further comparison we also report the outcome of the best performing probabilistic priority rule, respectively. To gain insight into the capabilities of the different algorithms, it is important to recall that problem 1 and problem 2 consist of less than one third of the operations to be scheduled in the other four test problems. In the following discussion we therefore distinguish between "small" and "large" problems.

Let us first have a look at the results achieved by Fang et al. Regarding T_n and T_{max}, their approach outperforms the best performing rule for every instance. Considering the mean measures $\overline{F}$ and $\overline{T}$, however, only the small problems are improved. For the large problems the approach is always inferior to the best performing rule. Therefore we conclude that pursuing a mean measure in job shop scheduling is a much more difficult optimization problem in general. For the small problems it is reported that the best schedules of mean measures have been obtained by the stochastic hill-climbing algorithm. This indicates that the GA of Fang et al. is basically unable to solve complex dynamic instances.

Considering the results reported by Lin et al., it is seen that small problems are solved by nearly the same solution quality already achieved by Fang et al. Notice that for small problems the best-known solutions have been generated either by one the or the other approach with respect to all criteria. Thus we suppose that small problems are solved to near-optimality by all approaches under consideration. Different than the small problems, the large problems are improved significantly by the GA of Lin et al. in every case. Hence, using advanced evolutionary techniques appears worth the higher runtime spent.

Nevertheless, the computation costs caused by the GA of Fang et al. are on average thirty times the costs of the method used throughout this chapter. Considering this enormous gap, it strikes that our GA produces a superior schedule quality in almost every case. The small problems are solved consistently within the same range of quality also obtained by all other approaches. Turning to the large problems it is verified that a substantial progress is still possible. Notice that the quality of schedules improves in particular for the difficult mean performance criteria $\overline{F}$ and $\overline{T}$. Since the presented adaptive search algorithm is also substantially faster, it is proved by comparison as a highly efficient method for dynamic scheduling.

3. SUMMARY

In this chapter we have dealt with some typical aspects of realistic scheduling scenarios. Additionally we have focussed on various objectives for planning. Unfortunately, neighborhood search cannot be applied to such problems because of a lack of efficient neighborhood definitions. Practice therefore relies on priority rules for capacity allocation. Against these simple but powerful rivals, adaptive search turned out consistently superior by experiment as well as pleasantly robust.

Comparing different schedule-decoding procedures, it is shown that the capabilities of GAs to conduct a proper search vanish with an increasing problem size. This weakness can be partly alleviated by adjusting the scope of search. For this end a tunable schedule decoding procedure has been proposed which proportions the size of a search space within useful bounds. Although this technique may exclude optimal and even near optimal solutions from being considered, the positive effects of reducing the search space do prevail. Another impact is a strong reduction of the computation costs. A short runtime in highly appreciated in practice because scheduling of activities is usually performed on a rolling time basis. If engaged for the rescheduling of activities, a further reduction of runtime is possible by falling back on information available at no cost. Here the GA population acts like an adaptive memory which is modified according to changes observed in the real-world process. By taking an overview of all results, it can be claimed that traditional rule-based algorithms for scheduling and rescheduling can be outperformed clearly by adaptive search at competitive costs.

Chapter 9

ADAPTIVE AGENTS AT WORK

Evolutionary search methods enable a thorough allocation of recourses with respect to physical constraints, individual priorities and business objectives. If such an approach is used in a changing environment, adaptation is able to respond adequately to unforeseen events. In this way the planning of day-to-day logistics operation is done reactively. Similarly, a person-machine dialog can intervene into the resource planning in order to generate different plans, to explore scenarios or to modify the constraints. Due to the adaptive memory maintained by EAs, the search process may be suspended and reoriented, but it can be continued afterwards on the basis of the previously gathered knowledge. In this context we view an EA as a permanent, reactive, goal-oriented learning entity, or simply as an adaptive agent.

This chapter presents two applications of adaptive scheduling agents. In the first section we consider a problem of scheduling customer jobs in an existing firm. The objective pursues just-in-time completion of final products at minimal production costs incurring from the deployment of staff. Since the manufacturing processes are embedded in a real-world environment, typical failures such as delayed material supplies and machine breakdowns are taken into a detailed consideration. The second investigation on adaptive agents addresses their application to control problems. Usually, simple priority rules are employed in this field because decisions have to be made on-line. By the use of adaptive agents, the runtime becomes the most crucial aspect of performance. Therefore we take advantage from adaptive memory access and show by experiment that adaptive agents enable a better system control than priority rules, even in heavy loaded situations.

1. A CASE STUDY FROM INDUSTRY

Modern concepts of industrial production like *Lean* and *Just-in-Time* have shown a world-wide impact in the last decade. One of the basic ideas emphasizes on-time delivery of sub-assemblies and final products. The transfer of this claim towards an operational implementation often remains an open issue because in many sectors of industry technical and organizational obstacles prevent an easy realization of challenging management concepts.

This section shows a transformation of the strategic aim of the just-in-time philosophy towards a practical objective for production planning. For an existing firm we design an adaptive agent capable to solve the difficult combinatorial problem arising in the logistics operation. A performance comparison of this automated planning approach with the former manual way of production scheduling is outlined by an extract of jobs actually processed in the workcenter of the firm within one calendar year.

1.1 THE PRODUCTION ENVIRONMENT

In the following we consider a medium-sized firm in the fastening sector, compare also Rixen et al. (1995); Rixen (1997). It produces specialized fastening tools for the motor-vehicle industry as well as different kinds of fasteners such as nails, staples and glue. In particular we study the scheduling of so called collator machines which connect certain types of nails to strips of collated nails fitting professional nailing tools. Jobs are given to a collator by specifying a lot-size and a certain nail type. The processing of different jobs may require changeover times which are regarded as a setup for the collator.

In order to serve the customer orders, the sell department passes on the demand to the manufacturing workcenter. On-time delivery of final products is of great importance for the firm. Early and tardy deliveries are highly discouraged because they lead to penalty costs. For this reason the sell department sets a due-date for each job which includes a security time span between the intended job completion and the delivery date. The jobs are scheduled in the workcenter without taking costs into account, that is, they are scheduled with respect to the time windows only. As soon as a job reaches the workcenter its raw material is ordered. The material arrives at the workcenter not before two weeks after ordering. The projected arrival time of the required material is used as the ready-time of a job.

Tab. 9.1 shows the chronology of events for one collator machine in the year 1994. At first the backlog of work which has been released

Table 9.1. Job data of a collator machine.

job i	*type*	r_i	s_i	q_i	p_i	d_i
Event I at *3.1.94:* Backlog of work from previous year						
J01	C06	06.12.93	3	34	0.1606	13.12.93
J02	C02	13.01.94	3	1200	0.0592	26.01.94
J03a	C06	13.01.94	3	325	0.1606	09.02.94
J03b	C06	02.02.94	3	325	0.1606	09.02.94
J04	C06	02.02.94	3	40	0.1606	13.12.93
J05a	C03	02.02.94	3	1107	0.0638	23.02.94
J05b	C03	16.02.94	3	93	0.0638	23.02.94
J06a	C04	18.02.94	3	1672	0.0688	23.03.94
J06b	C04	11.03.94	3	528	0.0688	23.03.94
J07	C01	11.03.94	3	800	0.0839	06.04.94
Event II at *8.2.94:* Job releases						
J08a	C04	08.04.94	3	1496	0.0688	01.06.94
J08b	C04	14.04.94	3	176	0.0688	01.06.94
J08c	C04	11.05.94	3	528	0.0688	01.06.94
J09	C02	11.05.94	3	1130	0.0592	15.06.94
J10	C03	11.05.94	3	800	0.0638	29.06.94
Event III at *11.5.94:* Job releases						
J11	C05	18.05.94	3	500	0.0943	29.07.94
J12	C04	18.05.94	3	2200	0.0688	15.08.94
J13	C03	18.05.94	3	1570	0.0638	16.09.94
J14	C06	18.05.94	3	330	0.1606	30.09.94
J15	C00	26.05.94	14.5	10	0.1898	24.06.94
J16	C06	14.06.94	3	750	0.1606	20.07.94
Event IV at *11.7.94:* Job releases						
J17	C00	20.07.94	14.5	50	0.1898	27.07.94
J18	C01	22.07.94	3	680	0.0839	28.09.94
J19	C06	22.07.94	3	620	0.1606	19.10.94
J20	C04	22.07.94	3	2570	0.0688	23.11.94
J21	C02	22.07.94	3	1030	0.0592	30.11.94
J22	C03	22.07.94	3	740	0.0638	21.12.94
Event V at *16.8.94:* Machine breakdown until 5.9.94						
Event VI at *28.9.94:* Job release						
J23	C05	12.10.94	3	500	0.0943	21.12.94

but not finished in 1993 is transferred. Most of these jobs have ready-times r_i within 1994, i.e. their material demand is not available within 1993. As shown in the table, different nail types can be processed by the collator machine. The setup s_i of the collator takes about three time units (measured in industrial working hours) if processing changes over to another nail type. Sometimes, several jobs of the same type are released at the same time, e.g. J03a, J03b. This is caused by split arrivals of raw material which enforce different ready-times for these jobs. A setup of the collator is of course unnecessary if two jobs of the same type are processed right after each other. The processing time of a job is obtained from the lot-size q_i and the duration p_i which specifies the time needed to process the least possible unit of a particular type of nail.

From time to time, special jobs are released by the research department of the firm. These prototype jobs are processed for innovative purpose only. Although the research department declares a due-date for these jobs, they have no definite date of delivery. Tab. 9.1 shows the prototype jobs J15 and J17, both characterized by a long setup and a small lot-size.

We now describe how machine schedules are actually built in the workcenter. Fastener products are usually produced by a single dedicated machine. The workcenter is therefore confronted with several independent one-machine scheduling problems of the type described above. For each machine a new schedule which contains all released and not yet started jobs is built every two months. The building of a new schedule is triggered by the desire to estimate the utilization of machines within the near future. Next to that it can be forced by other events, for instance a sudden machine breakdown. Relying on expert knowledge rather than optimization methods, the machine schedules are still made by hand in the workcenter. If e.g. a job risks delay its completion may be accelerated by using double shifts. Nevertheless, wage-intensive double shifts are introduced in urgent situations only.

To summarize the production environment of the considered firm, jobs are processed by single machines without preemption. They are released at undetermined points in time which actually occur rarely. There is a changeover time between jobs of different types. The primary measure of schedule performance responds to just-in-time completion of jobs. Secondary, double shifts can be performed but should be avoided whenever possible.

1.2 AN AUTOMATED SCHEDULING SYSTEM

In the following we propose an automated scheduling method for the workcenter which is based on the idea of adaptive agents. Starting point is the circumstance that scheduling is not yet event-driven in the workcenter. It is performed simultaneously for all machines about every two months because man-made scheduling of the large number of machines is very time consuming. In order to optimize the production processes on a day-by-day basis we suggest to control each machine by an individual agent. The agents sense the production environment and adapt the action of their machine to the subset of events which is relevant for that machine (e.g. a production interrupt or job releases). Thus we face a successive coordination of the machine activities as outlined in Chap. 2(3.2). Since the machines work independently of each other there is no communication necessary among the agents.

Tab. 9.1 shows a series of events for one collator machine. Whenever an event occurs the agent has to generate a new machine schedule, i.e. it determines starting times for new jobs and it reschedules starting times of the queuing jobs. The necessary modification of job data is done according to the techniques developed in the previous chapter. From the viewpoint of a single machine events occur rarely. The runtime efficiency of the agents is therefore of subordinate interest in favor of effectiveness. Incorporating the adaptive memory function of the agents is not needed. However, two things remain to be specified. This is (i) the evaluation of schedules according to a measure of just-in-time completion of jobs and (ii) the possible acceleration of urgent jobs by double shifts. Both aspects are subject of the next section.

1.3 DESIGN OF ADAPTIVE AGENTS

In order to depict the complete design of adaptive agents for the considered application, we refer to the architectural model developed in Chap. 5(3.). Merely three components of a scheduling agent need further description: the encoding-decoding scheme, the evaluation function, and finally the algorithm employed for discovering new solutions.

ENCODING SCHEDULES

We are looking for a schedule representation which depicts the considered one-machine scheduling problem right down to its important decision details. As outlined above, scheduling a machine requires two kinds of decisions in the workcenter. A schedule is constructed (i) with respect to the processing sequence of jobs and (ii) by means of some integers which represent the number of double shifts projected for the jobs. Other variables such as completion times, earliness, and tardiness of jobs should be expressed in terms of these both decisions. For this reason we propose a schedule representation of two interacting strings which is shown in Fig.9.1.

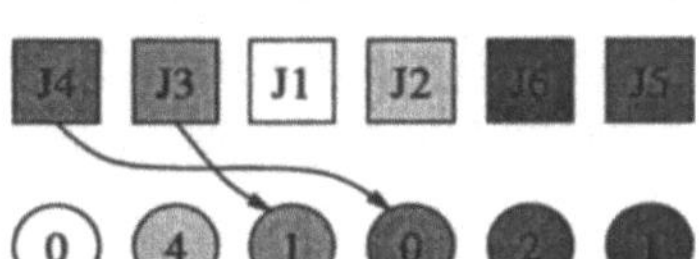

Figure 9.1 Encoding of job sequences and double shifts.

String 1 encodes a processing sequence for the jobs by our conventional permutation scheme. By this, a natural representation is given which covers each possible job sequence exactly once. String 2 encodes the number of double shifts used to process each of the jobs. As indicated by the two arcs, the job identifier given in string 1 is used as an index pointing into string 2. In this way the number of double shift projected for a job is determined. In the example, J4 is processed first without double shift, followed by J3 using one double shift, etc. Notice that string 2 is handled as an integer-valued vector.

The integers carried in string 2 can range from zero to the largest possible number of double shifts of the corresponding job. For each job this number is bounded by the integer part of half of the entire processing time[1]. The use of integers is motivated by the idea that a double shift is introduced only if the capacity of this shift is fully exploited. For a value of zero a job will be processed without double shifts. Consider e.g. job J07 listed in Tab. 9.1. It occupies the collator by a total of $800 \cdot 0.0839 = 67.17$ industrial working hours. A single shift counts 7.5 industrial working hours. The job J07 can therefore be processed by inserting at most four double shifts which leads to an entire processing of approximately 5 working days. Without any double shift it occupies the collator for about 9 days.

DECODING SCHEDULES

The objective pronounced by Just-in-Time leads to a function of minimizing the mean absolute lateness, i.e. the sum of earliness and tardiness of the jobs[2]. In the workcenter under consideration this objective is pursued with respect to a minimal number of double shifts. Thus we are actually confronted with a multi-criteria optimization problem.

Presupposed that job sequences and double shifts are encoded in the problem representation, we have to determine reasonable job starting-times from this data. Unfortunately, the mean absolute lateness is no regular measure of schedule performance. Starting a job as early as possible effects that this job is completed too early in general. This means that semi-active scheduling works badly at least for jobs which are not time-critical. The schedule-builder of Chap. 8(1.5) cannot be used in this case.

Building just-in-time schedules therefore attempts to delay the start of the jobs as long as they can meet their desired completion time exactly. Machine idle times are inserted between the processing of consecutive

---

[1]Initial instantiations of string 2 are therefore randomly generated using a uniform integer distribution over the interval $[0, (q_i p_i/2]$ for J_i.

[2]A review of this measure of schedule performance is given by Fry et al. (1990). Already in the one-machine case the problem turns out being *NP*-hard. Fry et al. therefore propose a linear programming approach which is combined with a local-search heuristic based on the critical transpose neighborhood.

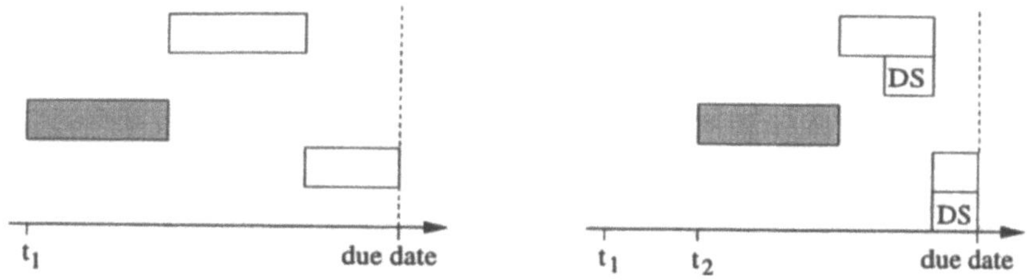

Figure 9.2. Scheduling of three jobs with identical due-dates. It is assumed that the grey job can be started at t_1 (left). Keeping the machine idle for a while this job can be alternatively started at t_2 (right) without delaying the subsequent jobs if double shifts are used to process these jobs.

jobs if this enhances the just-in-time measure. However, for the problem under consideration this strategy leads to an increasing number of double shifts if jobs have a similar due-date, see Fig. 9.2.

Although the grey job is less early in the right chart it enforces double shifts for the following jobs not to become tardy. Double shifts, however, are discouraged because they increase the costs of a working hour by about 20%. Since costs arising from job earliness cannot be quantified by the workcenter the number of double shifts is reduced with higher priority. In other words, we refrain from introducing idle times to the collator machine as sketched in Fig. 9.2. We suggest to build left-shifted schedules from the solution encoding instead. Now, the starting time t_i of job J_i is calculated iteratively from the starting time of the preceding job J_j in the permutation string by

$$t_i = \max(r_i\,,\, t_j + q_j p_j - N_j h + s_i). \tag{9.1}$$

Here, N_i denotes the number of double shifts used to process J_i and h=7.5 denotes the number of industrial working hours per shift. Hence, $N_j h$ determines the total number of costly working hours which are planned to accelerate the preceding job J_j. Notice that this time span has to be subtracted from the regular completion of the predecessor whereas the changeover time is added (if required) in order to determine the starting time for a job.

EVALUATION OF SCHEDULES

Given t_i and N_i projected for a job, its completion time is determined by $c_i = t_i + q_i p_i - N_i h$. Next, the earliness and tardiness is calculated by $E_i = \max(d_i - c_i, 0)$ and $T_i = \max(c_i - d_i, 0)$, respectively. For some jobs it may be very important not to be late, some other jobs should not be completed too early. In order to express a preference between early or tardy completion of a job, a specific weight ω_j $(0<\omega_j<1)$ can be used. The just-in-time performance of a job is measured then by

$$JIT_i = \omega_j E_i + (1-\omega_j)T_i. \tag{9.2}$$

If no preferences are given to a job the weight is set to $\omega_i = 0.5$[3]. Independent of the particular setting of ω_i, we can state a job to be completed just-in-time if and only if $JIT_i=0$. If n jobs are released for a machine, the overall scheduling objective is stated as to minimize

$$\overline{JIT} = \frac{1}{n}\sum_{i=1}^{n} JIT_i. \tag{9.3}$$

For prototype jobs the on-time measure is often rather unimportant. In order to ensure that these jobs are always processed without double shifts, we can set their weight to $\omega_j = 1$ beforehand. This schedule evaluation realizes the concept of just-in-time production with respect to the details of the manufacturing workcenter considered. Of course, some important details are hidden in the problem encoding and therefore we finally have to concentrate on the operators used for recombination.

SCHEDULE DISCOVERY

Recall that the possible schedules are encoded by two strings, one representing the processing sequence of jobs and the other one the corresponding numbers of double shifts used to process the jobs. Suitable recombination operators are briefly discussed for both types of strings.

Since an agent is confronted with only one machine its search mechanism should work most respectful concerning the absolute ordering of the jobs. Consider a job which has to be processed early because of an early due-date. After a crossover this job should progressively appear somewhere at the beginning of string 1. Thus we could use the PMX operator which effects the least positional bias among the three permutation crossover techniques compared in Tab. 6.2. Even more reliable with respect to this characteristic works *position-based crossover* as proposed by Syswerda (1991). It is therefore employed for the crossover operation. By the same argumentation we use the pairwise-exchange operator for mutating string 1, compare Tab. 6.1.

The second string responds to the vector type. Two-point crossover performs a valid syntactical operation then. Mutations are simply effected by choosing a field of the vector at random before the addressed integer is slightly modified with respect to the feasible bounds.

---

[3]Notice that this setting implicitly tends to prefer a shorter tardiness at the expense of a shorter earliness because schedules are builded in left-shifted manner.

1.4 SIMULATION STUDY

Following the job data in Tab. 9.1 we are going to simulate the behavior of the adaptive agent. Its GA-based knowledge-discovery system is parameterized by a standard setting.

1. The population size is set to 128 individuals.
2. Scaled fitness-proportionate selection is used.
3. The recombination is performed as described above with a crossover probability of 0.5 and a mutation rate of 0.01.
4. The algorithm terminates after 100 iterations.

A balanced weight $\omega = 0.5$ is used for assessing the on-time quality of the non-prototype jobs by equation (9.2). The simulation of the agent is repeated for 30 times. Since the agent showed an extremely robust performance we only report the best generated machine schedule, see Tab. 9.2.

The jobs listed in the first column appear in the suggested processing order. Each event is reported by a subtable which shows the best schedule contained in the knowledge base of the agent at that point in time. The three columns of the subtables report the scheduled job completions, machine setups, and the intended number of double shifts for the jobs. A symbol "x" denotes that a setup is required. All dates are given in accumulated units of working hours. The entry date refers to the starting time $t = 0$ of the simulation. In order to avoid negative values of time, the due-date of the two jobs J01 and J04 (falling into 1993) is set to zero.

It can be seen from the table that most of the jobs are rescheduled several times. Consider e.g. jobs J06a and J07b. In the first period they are scheduled for completion at 427 and 405, respectively. But in the following period the processing order of the jobs is in reverse as shown by the new completion times of 404 and 496. Within this period both jobs are actually processed, they do not appear in the back part of the table again.

Of special importance is the behavior of the agent in case of a machine breakdown. Such an event is observed at a simulation time of $t = 1191$. The breakdown occurred while setting up the collator for J13. Afterwards the machine has been repaired and could not be used for 20 days. Fortunately, a two week vacation in the firm (ignored in the simulation time) caused a downtime of only 32 working hours. Notice that the processing order of the following seven jobs did not change after the downtime, but the number of double shifts increased strongly. This

Table 9.2. An agent-generated schedule for the collator problem of Tab. 9.1.

Job	Due	Event I			II			III			IV			V			VI			Just-in-Time	
id	date	C_j	s	N_j																E_j	T_j
J01	0	9	x	0																	9
J02	127	131	x	0																	4
J04	0	171	x	0																	171
J03b	202	200		3																2	
J03a	202	238		2																	36
J05a	277	281	x	4	281	x	4														4
J05b	277	287		0	287		0														10
J06b	427	427	x	2	404	x	0													23	
J06a	427	405		0	496		3														69
J07	495	497	x	0	567	x	0														72
J08a	780				673	x	0													107	
J10	930				847	x	0	734	x	0										196	
J08c	780				781	x	1	773	x	0										7	
J08b	780				793		0	785		0											5
J09	855				749	x	0	855	x	0										*	*
J15	915							871	x	0										44	
J16	1042							995	x	0										47	
J11	1095							1045	x	0	1045	x	0							50	
J17	1080										1057	x	0							23	
J12	1185							1184	x	2	1189	x	3								4
J13	1290							1287	x	0	1272	x	0	1296	x	4					6
J18	1350										1338	x	2	1348	x	1				2	
J14	1365							1343	x	0	1371	x	3	1382	x	3					17
J19	1455										1456		2	1459		3	1454		3	1	
J20	1635										1635	x	0	1639	x	0	1639	x	0		4
J21	1672										1681	x	2	1703	x	0	1703	x	0		31
J22	1785										1735	x	0	1752	x	0	1753	x	0	32	
J23	1785																1788	x	2		3
double shifts:				5			7			0			3			8			5	534	445

verifies that the agent adapts well to the stressed situation caused by a machine breakdown.

Let us finally have a look at the overall performance of the agent. The bottom row of Tab. 9.2 accumulates the double shifts used within the periods. Consider job J08c which is scheduled in the second period using one double shift and rescheduled in the third period again. Since this job did not start in the second period we count a total of only seven double shift instead of eight as originally intended. The right column of the table shows the just-in-time performance of every job in terms of earliness and tardiness. The only job which perfectly meets its due-date is J09. For most of the other jobs an acceptable deviation of a few hours is obtained. At least some jobs are clearly delayed. Consider e.g. J04 which has a large delay of 171 hours. Since its due-date hits the beginning of the simulation one may ask why the agent permits J02 to precede J04? The answer is given by the ready-times of both jobs in Tab. 9.2. The needed raw material for J04 does not reach the workcenter before February, so it is impossible to complete it on-time.

To summarize, we count 534 hours of earliness and a slight smaller number of 445 hours of tardiness. These values are achieved by inserting a total of 28 double shifts and lead to the just-in-time measure of $\overline{JIT}=17.5$. In comparison, the manual schedule really implemented on the collator in 1994 was assessed with a corresponding measure of $\overline{JIT}=33.3$. It required 38 double shifts at the same time.

1.5 DISCUSSION

In this section we have developed an automated job scheduling system for a workcenter of a nailing-tool firm. The promising simulation result reveals the amount of machine and staff capacity which is still unexploited. These potentials undoubtedly have to be utilized in order to match the future challenges of new concepts in logistics.

In the manufacturing workcenter the desired goal is approached by a good compromise between interdependent decisions. Under increasingly faster changing conditions such solutions cannot be induced by human experts anymore. The technology of adaptive agents shows a promising way to support daily operations in effective manner. From an academic point of view the constraints and the size of the considered problem appear to be still moderate, an expert however can hardly improve the solution obtained.

2. ON-LINE SYSTEMS

Due to recent advances in information and transportation technology, topical problems of planning the logistics operation receive progressive importance. Relevant fields address e.g. the automatic handling of material and the real-time scheduling of transport activities[4].

The great difficulty of on-line decision making is based on two obvious reasons. Whenever a running process demands an instruction, a decision has to be made and its implementation cannot be postponed. Hence there is not much time left for anticipating the consequences of potential decisions which, of course, is necessary in order to weigh up the alternatives. Since a thorough planning of activities is hindered, dispatching rules are often used in practice in order to control the current logistic processes[5]. If an adaptive agent is engaged for an on-line control task, its reaction time becomes a crucial aspect of performance. This section investigates the suitability of adaptive agents for on-line control problems by a simulation study in the field of production scheduling.

2.1 PRODUCTION CONTROL

With the spread of automatic manufacturing systems the problem of assigning jobs to machines dynamically has received increasing attention[6]. Due to the lack of adequate optimization methods, control approaches based on priority rules have a long tradition in production research[7].

The immanent weakness of priority rules is to effect state changes of a system without anticipating the consequences. Recent production research therefore concentrates on decentralized approaches where the aggregates of a manufacturing systems can exchange messages in order to decide on the particular rule to be applied in a certain situation[8]. The choice of an appropriate rule depends on the goal currently pursued, e.g. to reduce the workload, to supply an idle machine, to accelerate important jobs which risk to become tardy and so on.

Another approach to the control problem of automated manufacturing has been proposed by Raman et al. (1989). Based on a temporal decomposition of the non-deterministic problem, they consecutively apply a branch-and-bound method in order to assign jobs to machines dynamically. Of course, under real-time conditions this idea is computational

[4]cf. Ascheuer et al. (1999).
[5]e.g. First-Fit, Nearest-Neighbor, First-Comes First-Serves.
[6]cf. Van Dyke Parunak (1992).
[7]for comprehensive overviews see Baker (1974); Panwalkar and Iskander (1977); Blackstone et al. (1982) and Haupt (1989).
[8]cf. Zelewski (1993); Holthaus and Ziegler (1997).

prohibitive, but it opens at least the domain for optimization methods. Recently the idea received attention by research in evolutionary computation. The transfer of GA experiences to production control was firstly approached by Bierwirth et al. (1995) and has been taken up by Lin et al. (1997) and Rixen (1997). These approaches mainly address the GA effectiveness although the GA efficiency is certainly of predominant importance. So far however, the large runtime of adaptive scheduling has hindered a thorough computational study as usually done in order to assess the quality of priority rules.

In the following we deal with the application of adaptive agents to production control problems. For this purpose we run the GA developed throughout Chant. 8 inside of the agent framework. To accelerate the reaction time of the agent, state changes of the manufacturing system are monitored by its integral memory component[9]. In order to reveal the potentials of this approach for on-line control problems, a manufacturing system is simulated under variable utilization rates. The objective of the agent is to minimize the mean flow-time of jobs. The robustness of the GA, verified in Chap. 8(1.4), enables a straight-forward application also to other measures of performance.

For our purpose its sufficient to concentrate on the mean flow-time criterion of jobs, defined by equation (8.1). Furthermore, job-specific weights are neglected by assigning an identical weight to all jobs. As already verified in Chap. 8(1.3), the performance of every priority rule depends strongly on the objective pursued. Fortunately, for the classical objective of flow-time reduction there exists a dominating rule, namely SPT[10]. Thus we can compare the performance of the adaptive agent with a simple but very powerful rival.

2.2 MODEL OF MANUFACTURING SYSTEMS

The *workload* of a manufacturing system is regarded as a crucial aspect of its overall performance. Recall from Chap. 8(1.2) that the workload is defined by the number of operations in the system which await processing. The workload observed in a manufacturing system is influenced by two independent forces, the *arrival process* of incoming jobs and the scheduling authority which counteracts this process. Advantageous scheduling therefore accelerates the job flow-times in order to keep the workload within useful bounds.

[9] compare Fig. 8.4.

[10] cf. Haupt (1989).

We consider a manufacturing system which is organized as job shop. To observe the system under different workload conditions we refer to a simulation environment which is often used to evaluate priority rules[11].

1. The manufacturing system consists of $m=6$ machines.
2. Each job passes 4 to 6 machines resulting in 5 operations on average.
3. The machine routings of jobs are generated from a uniform distribution.
4. The processing times of operations are uniformly distributed in the range of [1, 19]. This leads to a mean processing time of $\overline{P} = 5 \cdot 10$.
5. We generate exponentially distributed inter-arrival times[12] of jobs by using utilization rates of $U \in \{0.65, 0.7, 0.75, 0.8, 0.85, 0.9\}$.

The utilization rate approximates a workload situation by neglecting the influence of scheduling. A utilization of $U \geq 1$ leads to an excessive workload which cannot be handled anymore. The rates used in the following experiments are considerable lower. Modeling the inter-arrival times by a Possion process as done above can lead to extreme deviations of the workload in different phases of a simulation run. Often a large variance of the job flow-times can be observed. In order to alleviate this effect, a large number of job releases has to be simulated. Moreover, the simulation has to repeated several times in order to produce a sound overall mean. For each utilization rate we therefore conduct 50 simulation runs consisting of 1 000 jobs each.

In a first study we carry out SPT runs for all utilization rates prescribed above. The workload observed in the system is recorded, which is shown in Fig. 9.3. One can easily recognize the six levels of workload corresponding to the six utilization rates used. Consider the development of the workload in the early phase of the simulation. It continuously increases until a stationarity is reached (apart from $U = 0.90$) after approximately 200 job releases. Notice that the difference between adjacent levels of stable workload increases. This observation indicates that a small additional increase of the workload may not be handled properly in a tense production situation anymore. For this reason SPT-based control is unable to stabilize the workload for a machine utilization of $U = 0.90$ within the 1000 jobs released.

The figure depicts that jobs scheduled in the early phase of the simulation distort the mean flow-time measure considerably. The same is

[11] cf. Holthaus and Rajendran (1997).

[12] with a mean of $\lambda = \overline{P}/m\,U$, compare equation (8.7).

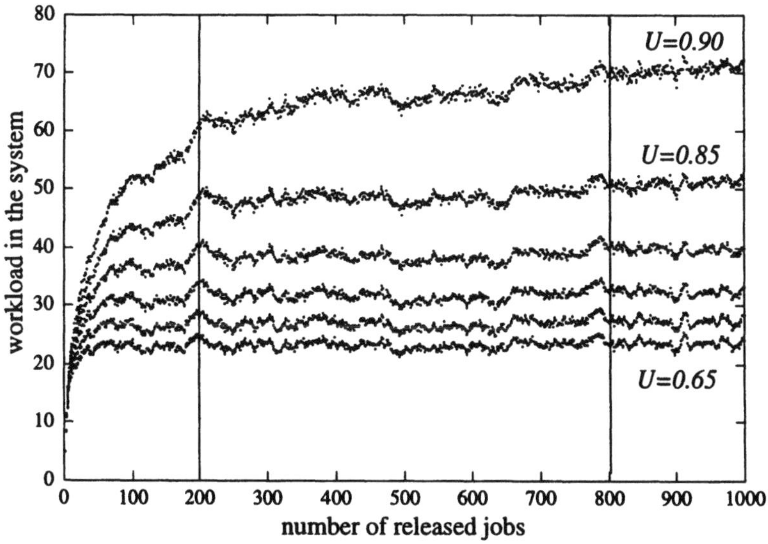

Figure 9.3. Workload under SPT-based control of a manufacturing system.

true for jobs released in the final phase of a simulation run. Therefore we discard job 1 to 200 as well as job 801 to 1,000 from being evaluated in order to circumvent distortion effects.

2.3 DESIGN OF ADAPTIVE AGENTS

We run the GA previously developed to solve machine scheduling problems by only slight modifications inside of an adaptive agent. We can think of this agent perceiving its environment as sketched in Fig. 8.4. It provides a schedule for all jobs currently known in the manufacturing system at any time. The learning algorithm used by the agent continuously searches for a better schedule regarding the mean flow-time of jobs. The parameters of this GA are adopted from the setting used in Chap. 8(1.4).

1. The population size is set to 100 individuals.
2. Scaled fitness-proportionate selection is used.
3. The recombination calls PPX crossover at probability of 0.8 in combination with *random reinsertion* mutations at probability of 0.2.
4. The algorithm terminates after τ iterations are carried without gaining any further improvement. The parameter τ is set to the number of jobs contained in the current scheduling problem.

The crossover operation is applied at a slightly higher rate in order to speed-up the convergence. To take advantage from a fast conver-

Table 9.3. Problem parameters for a typical mix of scenarios.

cluster	easy		moderate		hard	
number of jobs	10	20	30	40	50	60
utilization rate	0.5	0.6	0.7	0.8	0.9	1.0

gence, a flexible termination criterion is used again. Notice that the previously used criterion was coupled with the number of operations contained in a problem. Using the smaller number of jobs instead, we intend to accelerate the agent's reaction times once more. Finally, a variable schedule-builder is engaged as a decoding procedure. Recall from Chap. 8(1.5) that its parameter δ allows to scale the search space of the GA towards an effective size. Since the size of the search space has also a strong influence on the runtime, we address the setting of δ by a computational study.

In order to find an appropriate scope of search by experiment, we refer to a mix of scenarios the agent might be confronted with during the simulation. In other words, we generate a heterogeneous testbed of scheduling problems which contains instances of typical size and complexity. For this purpose we vary the number of jobs between 10 and 60 in steps of 10. The utilization rate is varied between 0.5 and 1.0 in steps of 0.1. Altogether 300 test instances are generated. These instances are divided into three clusters of 100 instances each. The combinations of parameters used to generate the clusters "easy", "moderate" and "hard" are shown in Tab. 9.3. For each combination of parameters belonging to one cluster, 25 instances are generated.

"Easy" instances consists of a few jobs only which are released with loosely set inter-arrival times. "Hard" instances consists of many jobs which arrive tight after another. An increasing utilization rate effects a faster job release which in turn results in a larger number of operations to be scheduled simultaneously[13]. In short, a large utilization rate generates difficult problems and vice versa. In this way it is hoped to generate a typical mix of more or less challenging scenarios.

The above GA is run for all test instances by using the tunable schedule-builder with $\delta \in \{0, 0.1, \ldots, 1\}$. The mean flow-times of jobs gained by a particular value of δ are averaged over all instances that belong to the same cluster. Finally, the achieved results are normalized

[13]this is expressed formally by equation (8.6).

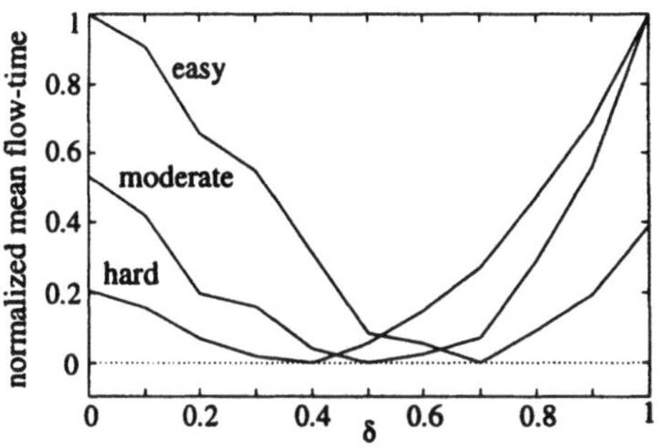

Figure 9.4 Influence of δ on the GA performance in the three clusters.

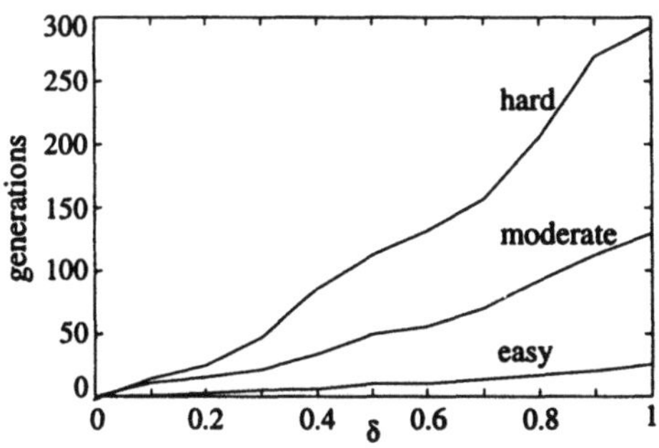

Figure 9.5 Influence of δ on the agent reaction-time in the three clusters.

into the range of [0, 1]. In this way they indicate a suitable parameterization of the schedule-builder.

This is shown in Fig. 9.4. Starting from non-delay scheduling with $\delta = 0$, the increase of δ leads always to improved solutions. Previously excluded solutions are now included into the scope of the search while the GA still works sufficiently well. Beyond a borderline a further increase of δ leads to a deterioration of the solution quality[14]. We find different values of δ to be most suitable in the three clusters, namely 0.4 for "easy", 0.5 for "moderate", and 0.7 for "hard" problem instances. One can easily identify that the worst level of solution quality for "easy" problems is produced with non-delay scheduling, while active scheduling still results in a reasonable solution quality. For the "hard" problem instances things are just the other way round. In case of extremely large problems a value of δ close to zero may work best.

We already known from Chap. 8(1.5) that a GA using an active schedule-builder needs more generations to converge than its non-delay counterpart. This is because it has to explore a larger search space. In order to reveal the impact of δ on the reaction time of the agent we now refer to an additional measurement taken in the above experiment.

[14]For an explanation and detailed discussion of this observation the reader is referred back to Chap. 8(1.5).

Fig. 9.5 depicts the development of the generations needed until the GA terminates. This presentation focuses on the average retardation of the GA by increasing the scope of the schedule-builder. Notice that the generations needed under $\delta=0$ are subtracted from the measurements. As one could expect, an approximately linear retardation of the GA is observed for increasing δ values in all clusters.

To summarize the experiment, a good trade-off between runtime and solution quality is observed at $\delta=0.5$ for all of the three clusters investigated. Here, the schedule-builder produces above average results by saving a significant amount of runtime at the same time. Particularly in situations of heavy workload, the expected loss of solution quality seems to be bounded. Therefore we parameterize the GA with $\delta=0.5$ in the following

2.4 SIMULATION STUDY

After we have carefully designed an adaptive agent for production control problems we examine it by a simulation study. In order to compare its performance with the outcome of SPT, we use the utilization rates as stated in Sect. 2.2. Rates of $U \in \{0.65, 0.70\}$ represent a relaxed situation for the manufacturing system. A moderate load is produced by $U \in \{0.75, 0.80\}$ whereas utilization rates of $U \in \{0.85, 0.90\}$ produce an excessive workload.

The production control agent (PCA) is tested in two variants. PCA^- is an ordinary adaptive agent which runs the described GA inside. Its counterpart additionally uses the integral memory component. The adaptive memory access of PCA^+ can have two opposite impacts on the quality of control. The information preserved in the GA population may concentrate the internal search process on parts of the search space which are still promising in a changed situation. On the other hand, the lower diversity in the GA population may also hinder a thorough search. Anyway, the faster convergence of the learning process for PCA^+ will certainly result in considerably shorter reaction times.

The simulation of PCA^- and PCA^+ is carried out like done for SPT, i.e. we conduct 50 simulation runs consisting of 1 000 jobs each for all utilization rates under consideration. Tab. 9.4 reports the average mean flow-time of jobs ($\overline{F}$) observed for SPT, PCA^- and PCA^+. For both agent approaches we additionally provide the average number of generations (gen.) needed to learn a changed environment and the average runtime (sec.) required for a single simulation run[15].

[15]measured on a Pentium/200 Mhz computer.

Table 9.4. Results of non-deterministic scheduling.

	SPT	PCA^-			PCA^+		
U	$\overline{F}$	$\overline{F}$	gen.	sec.	$\overline{F}$	gen.	sec.
0.65	91.8	85.8	10.8	111.7	85.9	8.8	81.1
0.70	100.4	93.2	14.9	189.6	93.5	11.1	127.6
0.75	111.5	103.5	21.4	351.0	103.1	14.6	214.6
0.80	128.1	118.7	33.5	747.6	117.4	20.3	407.5
0.85	154.9	142.6	56.7	1876.4	140.6	31.6	953.0
0.90	200.4	184.3	103.8	5928.1	179.4	52.4	2705.5

The table shows a strong increase of flow-times for an SPT-based control of the manufacturing system. Obviously, under higher utilization rates the workload results in longer job flow-times. The known average processing time $\overline{P} = 50$ of the jobs can serve as a simple lower bound for $\overline{F}$. By subtracting $\overline{P}$ from $\overline{F}$ we obtain the average waiting time of jobs. It increases from 42 time units for $U = 0.65$ up to 150 units for $U = 0.90$.

The SPT rule is outperformed clearly by both agents. A reduction of 15% of the waiting time of jobs is gained for all utilization rates. Yet, the control quality of both agents hardly differs from each other. As it can be taken from the lower curve in Fig. 9.3, the scenarios met for $U = 0.65$ consists of about 25 operations. Such small problems are solved reliably by both agents. With an increase of machine utilization the complexity of the scenarios increases as well and now PCA^+ dominates PCA^- by a few percent.

Tab. 9.4 further reports the reaction times of the agents measured by the number of generations carried out inside the GA. The use of adaptive memory access reduces the reaction times only slightly for $U = 0.65$. In contrast, for a heavily loaded system, PCA^- demands nearly double the generations of PCA^+. The larger the workload increases, the more advantageous adaptive memory access gets in terms of generations spent.

The benefit of adaptive memory access becomes even more clear if the simulation runtime is compared for both approaches. The percentage of runtime saving is actually larger than one may expect from comparing the generation columns in Tab. 9.4. This is verified by Fig. 9.6. The relative reduction of reaction time in seconds (observed for PCA^+ against PCA^-) is about 5% larger than the relative reduction of reaction time measured in terms of generations. Not only that less generations are spent for learning if adaptive memory is used, it also produces an additional value. The explanation of this effect is simple. Due to an decreas-

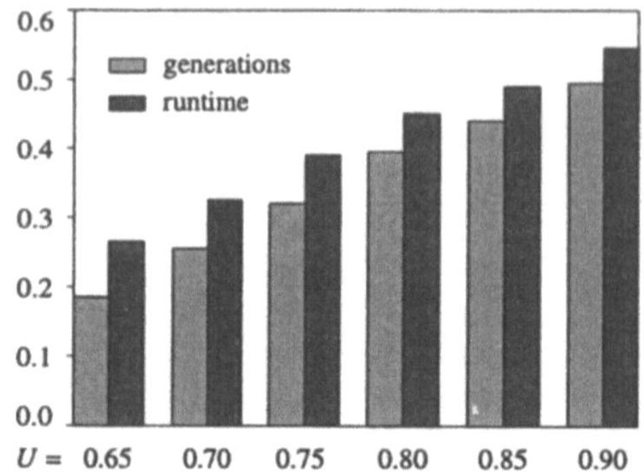

Figure 9.6
Relative decrease of generations needed vs. runtime savings gained from adaptive memory access.

ing diversity, the effort spent on evaluating a population decreases from generation to generation. Since PCA$^+$ starts already from a population of lower diversity, it profits from this general principle.

To summary, adaptive memory access speeds up the learning process of an agent considerably. It does not worsens its ability to control the manufacturing system on the other hand. Thus we conclude that PCA$^+$ is more efficient than PCA$^-$. For the highest machine utilization considered in our tests, the average workload can amount to more than 70 operations[16]. Even under this heavy load PCA$^+$ needs only 2705.5 sec. in order to cope with 1,000 incoming jobs. The average reaction time of 2.7 sec. appears competitively fast. The quality of an SPT-based control is outperformed by tidy 10% under such conditions, but we must admit that its reaction time requires a fraction of a second only.

2.5 DISCUSSION

Different to priority rules, adaptive agents can anticipate the consequences of their decisions to a certain extent. A major finding drawn from this study is that anticipation is useful, especially under heavy workload conditions. High utilization rates lead to long queues of jobs waiting in front of the machines. The longer a queue grows, the more potential decisions can be evaluated by an agent. Unfortunately, the advantage taken from anticipation is paid by a high computational effort. Long queues result in large decision problems which come along with considerable reaction times. Thus we state that the more attractive anticipation gets, the more expensive it is to achieve.

The use of an adaptive memory has shown a general way to alleviate the immanent weakness of on-line optimization methods. Nevertheless, if retardation of reactions can effect serious system failures, the donors

[16] see Fig. 9.3.

of optimization will hardly replace the behavior of simple dispatching rules. On the other hand, if already small improvements of control lead to considerable cost reductions, longer reaction times may be accepted. This motivates future research in methods combining anticipation with reactive behavior.

3. SUMMARY

This chapter has presented two applications of adaptive agents in the scheduling domain. While the emphasis is on planning rather than control in the first application it is the other way round in the second one. In both cases the agent approach is superior to the conventional way of problem-solving. Moreover, the profit gained by our effort is considerable. Outperforming e.g. SPT by ten percent is surprisingly much considering the fact how well this rule works for flow-time minimization. For objectives of higher economic relevance like tardiness minimization, the benefit will be even larger because there are no dominant rules anymore[17]. A mix of priority rules is often used in practice, but the aim to reliably adapt the logistic processes is hardly reached thereby. The proof of this claim is offered by our framework which combines simple components adopted from previous research in Operations Research and Evolutionary Computation into a powerful adaptive instrument.

[17] cf. Mattfeld and Bierwirth (1998).

Epilogue

Caused by the progressive competition in the logistics sector, anticipation combined with reactive behavior has become a business advantage for many industrial enterprises. While anticipation is fairly well developed by aggregate planning methods, short-term response on detailed demand is hardly possible, although reactive problem-solving goes for an inevitable hallmark of decision support in competitive logistics systems.

Throughout this work we have investigated a class of adaptive optimization methods which offer the required properties. Employing the simple yet powerful framework of evolutionary search allows to explore the space of logistics operation. Coupled with concepts of agent theory this framework permits a straight-forward extension of optimization tasks towards problem-solving in dynamically changing environments. The integral memory component of adaptive agents ensures that both, anticipation and reactive behavior, are brought together.

The concept has been proved by diverse computational experiments. We have chosen scheduling problems as one essential of detailed planning on the operation level because the profit to be gained from a better support is immense. This may warrant our focus, but even more the versatility of the developed representation scheme which enables a transfer to other problems. Thus it is hoped to glean the essentials of successful adaptive search for the management of logistics systems.

The interaction between autonomous systems develops a quite natural model of the interface between two actors of the logistic supply chain. This motivates the idea to study the behavior of two adaptive agents sharing parts of their individual environments. By externally changing the environment of a single agent, we have verified its capabilities of adapting to new goals and constraints. If agents share an environment their action effects internal changes. Adaptation may now lead to a reconciliation and optimization of the mutual logistics operation, but we leave this question as an open issue for future research.

References

Aarts, E., van Laarhoven, P.J.M., Lenstra, J.K., Ulder N.L.J. (1994): A Computational Study of Local Search Algorithms for Job Shop Scheduling. *ORSA Journal on Computing*, 6:18-125

Aarts, E., Lenstra, J.K., eds. (1997): *Local Search in Combinatorial Optimization.* John Wiley & Sons, Chichester et al.

Aarts, E., Lenstra, J.K. (1997): Introduction. In Aarts and Lenstra (eds.), pp. 1-17

Adams, J., Balas, E., Zawack, D. (1988): The Shifting Bottleneck Procedure for Job Shop Scheduling. *Management Science*, 34:391-401

Alander, J.T. (1998): An Indexed Bibliography of Genetic Algorithms in Logistics. Dept. of Information Technology and Production Economics, Report No. 94/1, University of Vaasa, Vaasa

Anderson, E.J., Glass, C.A., Potts, C.N. (1997): Machine Scheduling. In Aarts and Lenstra (eds.), pp. 361-414

Angeline, P.J. (1995): Adaptive and Self-Adaptive Evolutionary Computation. In Palaniswami et al. (eds.), pp. 152-163

Angeline, P.J. (1997): Tracking Extrema in Dynamic Environments. In McDonald and Eberhart (eds.), pp. 335-345

Arrow, K.J. (1964): Control in Large Organizations. *Management Science*, 10:397-408

Ascheuer, N., Grötschel, M., Krumke, S.O., Rambau, J. (1999): Combinatorial Online Optimization. In Kall, P. Lüthi, H.-J. (eds.): *Operations Research Proceedings 1998.* Springer-Verlag, pp. 21-37

Axelrod, R. (1987): The Evolution of Strategies in the Iterated Prisoner Dilemma. In Davis (ed.), pp. 32-41

Bachem, A., Hochstätter, W., Malich, M: Simulated Trading - A new Approach for Solving Vehicle Routing Problems. Faculty for Mathematics, Report No. 125, Dept. of University of Cologne, Cologne

Bäck, T., ed. (1997): *Proceedings of the 7th Int. Conference on Genetic Algorithms.* Morgan Kaufmann Publishers, San Francisco CA

Bäck, T., Hammel, U., Schwefel, H.-P. (1997): Evolutionary Computation: Comments on the History and Current State. *IEEE Transactions on Evolutionary Computation*, 1:3-16

Baker, K.R. (1974): *Introduction to Sequencing and Scheduling.* John Wiley & Sons, New York et al.

Balas, E., Vazacopoulos, A. (1998): Guided Local Search with the Shifting Bottleneck for Job Shop Scheduling. *Management Science*, 44:262-275

Beasley, D., Bull, D.R., Martin, R.R. (1993): An Overview of Genetic Algorithms: Part 1, Fundamentals. *University Computing*, 15:58-69

Bellman, R. (1957): *Dynamic Programming.* Princeton University Press

Bellman, R., Zadeh, L.A. (1970): Decision-Making in a Fuzzy Environment. *Management Science*, 17:141-164

Below, R.K., Booker, L.B., eds. (1991): *Proceedings of the 4th Int. Conference on Genetic Algorithms.* Morgan Kaufmann Publishers, San Mateo CA

Bierwirth, C. (1993): *Flowshop Scheduling mit Parallelen Genetischen Algorithmen.* DUV Gabler-Vieweg-Westdeutscher Verlag, Wiesbaden

Bierwirth, C. (1995): A Generalized Permutation Approach to Job Shop Scheduling with Genetic Algorithms. *OR Spektrum*, 17:87-92

Bierwirth, C., Kopfer, H., Mattfeld, D.C., Rixen, I. (1995): Genetic Algorithm based Scheduling in a Dynamic Manufacturing Environment. *Proceedings of the 2nd IEEE Int. Conference on Evolutionary Computation.* IEEE Press, Perth, pp. 439-443

Bierwirth, C, Mattfeld, D.C., Kopfer, H. (1996): On Permutation Representations for Scheduling Problems. In Voigt et al. (eds.), pp. 310-318

Bierwirth, C., Mattfeld, D.C. (1999): Production Scheduling and Rescheduling with Genetic Algorithms. *Evolutionary Computation*, 7:1-17

Biethahn, J., Nissen, V., eds. (1995): *Evolutionary Algorithms in Management Applications.* Springer-Verlag, Berlin et al.

Blackstone J.H., Phillips D.T., Hogg G.L. (1982): A State-of-the-Art Survey of Dispatching Rules for Manufacturing Job Shop Operations. *Int. Journal of Production Research*, 20:27-45

Blanton, J.L., Wainwright, R.L. (1993): Multiple Vehicle Routing with Time and Capacity Constraints using Genetic Algorithms. In Forrest (ed.), pp. 452-459

Błażewicz, J., Domschke, W. and Pesch, E. (1996): Job Shop Scheduling Problems: Conventional and New Solution Techniques. *EJOR*, 93:1-30

Bowersox, D.J. (1999): Logistics: From Necessity to Competitive Advantage. *Logistik Management*, 1:35-39

Bramel, J., Simchi-Levi, D. (1993): *The Logic of Logistics.* Springer-Verlag, Berlin et al.

Bruns, R. (1993): Direct Chromosome Representation and Advanced Genetic Operators for Production Scheduling. In Forrest (ed.), pp. 352-359

Burkhard, R.E., Rendl, F. (1984): A Thermodynamically motivated Simulation Procedure for Combinatorial Optimization Problems. *EJOR* 17:169-174

Caruana, R.A., Schaffer, J.D. (1988): Representation of Hidden Bias: Gray vs. Binary Coding for Genetic Algorithms. In: *Proceedings of the 5th Int. Conference on Machine Learning.* Morgan Kaufmann Publishers, Los Altos CA

Caudell, T.P., Dolan, C.P. (1989): Parametric Connectivity: Training of Constrained Networks using Genetic Algorithms. In Schaffer (ed.), pp. 370-374

Clarke, G., Wright, J.W. (1964): Scheduling of Vehicles from a Central Depot to a Number of Delivery Points. *Operations Research*, 12:568-581

Cobb, H.G., Grefenstette, J.J. (1993): Genetic Algorithms for Tracking Changing Environments. In Forrest (ed.), pp. 523-530

Colorni, A., Dorigo, M., Maffioli, F., Maniezzo, V., Righini, G., Trubian, M. (1996): Heuristics from Nature for Hard Combinatorial Problems. *Int. Trans. on Operational Research*, 3:1-21

Cotta, C., Troya, M.J. (1998): Genetic Forma Recombination in Permutation Flowshop Problems. *Evolutionary Computation*, 6:25-44

Council of Logistics Management (1995): *World Class Logistics: The Challenge of Managing Continuous Change.* Oak Brook IL

Dauzère-Péres, S., Lassere, J.B. (1994): Integration of Lotsizing and Scheduling Decisions in a Job-Shop. *EJOR*, 75:413-426

Davidor, Y. (1991): A Naturally Occurring Nice & Species Phenomenon: The Model and First Results. In Below and Boocker (eds.), pp. 257-263

Davidor, Y, Schwefel, H.-P., Männer, R., eds. (1994): *Parallel Problem Solving from Nature III.* Springer-Verlag, Berlin et al.

Davis, L.D. (1985): Job Shop Scheduling with Genetic Algorithms. In Grefenstette (ed.), pp. 136-140

Davis, L.D., ed. (1987): *Genetic Algorithms and Simulated Annealing.* Pitman Publishing, London

Davis, L.D., ed. (1991): *Handbook of Genetic Algorithms.* Van Nostrand Reinhold, New York

Davis, L.D. (1991): A Genetic Algorithm Tutorial. In Davis (ed.), pp. 1-101

Dawid, H. (1996): *Adaptive Learning by Genetic Algorithms - Analytical Results and Applications to Economic Models.* Springer-Verlag, Berlin

De Jong, K. (1990): Genetic-Algorithm-based Learning. In Michalski and Kodratoff (eds.), pp. 611-638

De Jong, K. (1991): Genetic Algorithms are Not Function Optimizers. In Rawlins (ed.), pp. 5-17

Della Croce, F. D., Tadei, R., and Volta, G. (1995): A Genetic Algorithm for the Job Shop Problem. *Computers and Operations Research*, 22:15-24

Dell' Amico, M., Trubian, M. (1993): Applying Tabu Search to the Job Shop Scheduling Problem. *Annals of Operations Research*, 41:231-252

Dixon, P.S., Silver, E.A. (1981): A Heuristic Solution Procedure for the Multi-Item, Single-Level, Limited Capacity, Lot-Sizing Problem. *Journal of Operations Management*, 2:23-39

Dorigo, M., Maniezzo, V. (1993): Parallel Genetic Algorithms: Introduction and Overview of Current Research. In Stender, J. (ed.): *Parallel Genetic Algorithms: Theory & Applications.* IOS Press, Amsterdam

Dorndorf, U., Pesch, E. (1995): Evolution Based Learning in a Job Shop Scheduling Environment. *Computers & Operations Research*, 22:25-40

Dowsland, K.E. (1993): Simulated Annealing. In Reeves (ed.), pp. 20-69

Du, J., Leung, J.Y. (1990): Minimizing Total Tardiness on one Machine is *NP*-hard. *Mathematics of Operations Research*, 15:483-495

Dueck, G., Scheuer, T. (1990): Threshold Acceptance: A General Purpose Algorithm Appearing Superior to Simulated Annealing. *J. Computational Physics*, 90:161-175

Eberhart, R.C. (1995): Computational Intelligence: A Snapshot. In Palaniswami et al. (eds.), pp. 9-16

Eiben, A.E., Raué, P.E, Ruttkay, Z. (1994): Genetic Algorithms with Multi-Parent Recombination. In Davidor et al. (1994), pp. 78-87

Eshelman, L.J., Schaffer, J.D. (1991): Preventing Premature Convergence in Genetic Alg. by Preventing Incest. In Below and Boocker (eds.), pp. 115-122

Eshelman, L.J. ed. (1995): *Proceedings of the 5th Int. Conference on Genetic Algorithms.* Morgan Kaufmann Publishers, San Mateo CA

Fahl, B. (1995): *Einsatz Genetischer Algorithmen zur Losgrößenplanung.* Diploma Thesis, University of Bremen, Bremen

Falkenauer, E. (1995): Tapping the Full Power of Genetic Algorithms through Suitable Representation and Local Optimization: Application to Bin Packing. In Biethahn and Nissen (eds.), pp. 167-182

Falkenauer, E. (1996): A Hybrid Grouping Genetic Algorithm for Bin Packing. *Journal of Heuristics*, 2:5-30

Fang, H.-L., Ross, P., Corne, D. (1993): A Promising Genetic Algorithm Approach to Job-Shop Scheduling, Rescheduling, and Open-Shop Scheduling Problems. In Forrest (ed.), pp. 375-382

Fang, H.-L., Corne, D., Ross, P. (1996): A Genetic Algorithm for Job-Shop Problems with Various Schedule Quality Criteria. In Fogarty (ed.), *AISB Proceedings.* Springer-Verlag, Berlin et al., pp. 39-49

Fleurent, C., Ferland, J. (1994): Genetic Hybrids for the Quadratic Assignment Problem. *Discrete Mathematics and theor. Computer Science*, 16:173-188

Fogel, L.J., Owens, A., Walsh, M. (1966): *Artificial Intelligence Through Simulated Evolution.* John Wiley & Sons, New York

Forrest, S., ed. (1993): *Proceedings of the 5th Int. Conference on Genetic Algorithms.* Morgan Kaufmann Publishers, San Mateo CA

Fox, R.B., McMahon, M.B. (1991): Genetic Operators for Sequencing Problems. In Rawlins (ed.), pp. 284-300

Franklin, S., Graesser, A. (1997): Is it an Agent, or just a Program?: A Taxonomy for Autonomous Agents. In Müller, J., Wooldridge, M., Jennings, N.R. (eds.): *Agent Theories, Architectures, and Languages III.* Springer-Verlag, pp. 21-36

Freisleben, B., Merz, P. (1996): New Genetic Local Search Operators for the Traveling Salesman Problem. In Voigt et al. (eds.), pp. 890-899

French, S. (1982): *Sequencing and Scheduling.* Ellis Horwood Limited Publishers, Chichester

Fry, T.D., Armstrong, R.D., Rosen, L.D. (1990): Minimize Mean Absolute Lateness: A heuristic Solution. *Computers Operational Research*, 17:105-112

Garey, M.R., Johnson, D.S. (1979): *Computers and Intractability: A Guide to the Theory of NP-Completeness.* W.H. Freeman and Co., New York

Gendreau, M., Laporte, G., Potvin, JY. (1997): Vehicle Routing - Modern Heuristics. In Aarts and Lenstra (eds.), pp. 311-336

Geyer-Schulz, A. (1995): *Fuzzy Rule-Based Expert Systems and Genetic Machine Learning.* Physica Verlag, Heidelberg

Giffler, B., Thompson, G.L. (1960): Algorithms for Solving Production Scheduling Problems. *Operations Research*, 8:487-503

Glover, F. (1977): Heuristic for Integer Programming Using Surrogate Constraints. *Decision Science*, 8:156-166

Glover, F. (1989, 1990): Tabu Search Part I & II. *ORSA J. Computing*, 1:190-206, 2:4-32

Glover, F. (1995): Scatter Search and Star Paths: Beyond the Genetic Metaphor. *OR Spektrum*, 2/3:125-137

Goldberg, D.E., Richardson, J. (1987): Genetic Algorithms with Sharing for Multimodal Function Optimization. In Grefenstette (ed.), pp. 41-49

Goldberg, D.E., Segrest, P. (1987): Finite Markoc Chain Analysis of Genetic Algorithm. In Grefenstette (ed.), pp. 1-8

Goldberg, D.E., Smith, R.E. (1987): Nonstationary Function Optimization Using Genetic Dominance and Diploidy. In Grefenstette (ed.), pp. 59-68

Goldberg, D.E. (1989): *Genetic Algorithms in Search, Optimization and Machine Learning.* Addison Wesley, Reading MA et al.

Goldberg, D.E., Deb, K. (1991): A Comparative Analysis of Selection Schemes used in Genetic Algorithms. In Rawlins (ed.), pp. 69-93

Goodman, N. (1955): *Fact, Fiction and Forecast.* Bobb-Merrill, Indianapolis

Gorges-Schleuter, M. (1989): Asparagos: An Asynchronous Parallel Genetic Optimization Strategy. In Schaffer (ed.), pp. 422-427

Gorges-Schleuter, M. (1992): Comparison of Local Mating Strategies in Massively Parallel Genetic Algorithms. In Männer and Manderick (eds.), pp. 553-562

Grefenstette, J.J., ed. (1985): *Proceedings of the 1st Int. Conference on Genetic Algorithms.* Lawrence Erlbaum Associates, Hillsdale

Grefenstette, J.J., ed. (1987): *Proceedings of the 2nd Int. Conference on Genetic Algorithms.* Lawrence Erlbaum Associates, Hillsdale

Grefenstette, J.J., (1987): Incorporating Problem Specific Knowledge into Genetic Algorithms. In Davis (ed.), 42-60

Grefenstette, J.J. (1992): Genetic Algorithms for Changing Environments. In Männer and Manderick (eds.), pp. 137-144

Grefenstette, J.J., (1993): Deception Considered Harmful. In Whitley (ed.), pp. 75-91

Günther, H.-O. (1987): Planning Lot Sizes and Capacity Requirements in a Single Stage Production System. *EJOR*, 31:223-231

Guntram, U. (1985): Die allgemeine Systemtheorie - Ein Überblick. *ZfB*, 55:296-323

Harp, S.A., Samad, T., Guha, A. (1989): Towards the Genetic Synthesis of Neural Networks. In Schaffer (ed.), pp. 360-369

Harrald, P.G., Kamstra, M. (1997): Evolving Neural Networks to Combine Financial Forecasts. *IEEE Transactions on Evolutionary Computation*, 1:40-52

Haupt R (1989): A Survey of Priority Rule-Based Scheduling. *OR Spektrum*, 11:3-16

Hax, A.C., Meal, H.C. (1975): Hierarchical Integration of Production Planning and Scheduling. In Geissler M.A. (ed.): *Studies in Management Sciences.* Elsevier Publishers, New York, pp. 53-69

Hayes-Roth, B. (1995): An Architecture for Adaptive Intelligent Systems. *Artificial Intelligence*, 72:329-365

Helber, S. (1994): *Kapazitätsorientierte Losgrößenplanung in PPS-Systemen.* M&P Verlag für Wissenschaft und Forschung, Stuttgart

Hertz, J., Krogh, A., Palmer, R.G. (1991): *Introduction to the Theory of Neural Computation.* Addison Wesley, Reedwood City CA

Holland, J.H. (1975 and 1992): *Adaptation in Natural and Artificial Systems.* MIT Press, Cambridge MA

Holland, J.H., Holyaok, K.J., Nisbett, R.E., Thagard, P.R. (1986): *Induction - Processes of Inference, Learning, and Discovery.* MIT Press, Cambridge MA

Holland, J.H. (1995): *Hidden Order: How Adaptation builds Complexity.* Addison-Wesley, Reading MA et al.

Holthaus O. (1996): *Ablaufplanung bei Werkstattfertigung.* DUV Gabler-Vieweg-Westdeutscher Verlag, Wiesbaden

Holthaus, O., Ziegler, H.. (1997): Look Ahead Job Demanding for Improving Job Shop Performance. *OR Spektrum*, 19:23-29

Holthaus, O., Rajendran, C. (1997): Efficient Dispatching Rules for Scheduling in a Job Shop. *Int. Journal of Production Economics*, 48:87-105

Hofbauer, J., Sigmund, K.: *Evolutionstheorie und Dynamische Systeme.* Verlag Paul Parey, Berlin and Hamburg

Horn, J., Goldberg, D.E. (1995): Genetic Algorithm Difficulty and the Modality of Fitness Landscapes, In Whitley and Vose (eds.), pp. 241-269

Hynynen, J. (1988): *A Framework for Coordination in Distributed Production Management.* Acta Polytechnica Scandinavica, M 52, Helsinki

Inayoshi, H., Manderick, B. (1994): The Weighted Graph Bi-Partitioning Problem: A Look at Genetic Algorithm Performance. In Davidor et al. (1994), pp. 617-625

Jang, J.S.R, Sun, C.T., Mizutani, E. (1997): *Neuro-Fuzzy and Soft Computing: A Computational Approach to Learning and Machine Intelligence.* Prentice Hall, Upper Saddle River NJ

Johnson, D.S., McGeoch, L.A. (1997): The Traveling Salesman Problem: a Case Study. In Aarts and Lenstra (eds.), 215-310

Jones, T. (1995): One Operator, One Landscape. Working Paper, Santa Fe Institute, #95-02-025, 1995, Santa Fe

Jones, T., Forrest, S. (1995): Fitness Distance Correlation as a Measure of Problem Difficulty for Genetic Algorithms. In Eshelman (ed.), pp. 184-192

Kargupta, H., Deb, K., Goldberg, D.E. (1992): Ordering Genetic Algorithms and Deception. In Männer and Manderick (eds.), pp. 47-56

Karr, C.L. (1991): Design of an Fuzzy Logic Controller using a Genetic Algorithm. In Below and Boocker (eds.), pp. 450-457

Kauffman, S.A. (1993): *The Origins of Order.* Oxford University Press, New York

Kindervater, G.A.P., Savelsbergh, M.W.P. (1997): Vehicle Routing - Handling Edge Exchanges. In Aarts and Lenstra (eds.), pp. 337-360

Kirkpatrik, S., Gelatt, C.D., Vecchi, M.P. (1983): Optimization by Simulated Annealing. *Science*, 220:671-680

Kirkpatrick, S., Toulouse, G. (1985): Configuration Space Analysis for Traveling Salesman Problems. *J. de Physique*, 46:1277-1292

Kistner, K.-P. (1992): Koordinationsmechanismen in der Hierarchischen Planung. *ZfB*, 62:1125-1146

Kitano, H. (1990): Designing Neural Networks using Genetic Algorithms with Graph Generation System. *Complex Systems*, 4:461-476

König, W., Kurbel, K., Mertens, P., Pressmar, B., eds. (1996): *Distributed Information Systems in Business.* Springer-Verlag, Berlin et al.

Kopfer, H. (1989): *Heuristische Suche in Operation Research und Künstlicher Intelligenz.* Habilitation thesis, Free University of Berlin, Berlin

Kopfer, H., Pankratz, G., Erkens, E. (1994): Entwicklung eines hybriden Genetischen Algorithmus zur Tourenplanung. *OR Spektrum*, 16:21-31

Kopfer, H., Rixen, I., Bierwirth, C. (1995): Ansätze zur Integration Genetischer Algorithmen in der Produktionsplanung und -steuerung. *Wirtschaftsinformatik*, 37:571-580

Kopfer, H., Utecht, T. Bierwirth, C. (1996): Distributed Environments for Evolutionary Algorithms by means of Multi-Agent Applications. In König et al. (eds.), pp. 139-157

Koza, J.R. (1992): *Genetic Programming.* MIT Press, Cambridge MA

Van Laarhoven, P.J.M., Aarts, E.H.L., Lenstra, J.K. (1992): Job Shop Scheduling by Simulated Annealing. *Operations Research*, 40:113-125

Lee, M., Takagi, H. (1993): Integrating Design Stages of Fuzzy Systems using Genetic Algorithms. In: *Proceedings of the 2nd IEEE Int. Conference on Fuzzy Systems.* IEEE Press, San Francisco CA, pp. 612-617

Liepins, G.E., Hilliard, M.R.: Genetic Algorithms: Foundations and Applications. *Annals of Operations Research*, 21:31-58

Lin, S., Goodman, E., Punch, W. (1997): A Genetic Algorithm Approach to Dynamic Job Shop Scheduling Problems. In Bäck (ed.), pp. 481-489

Lin, S. (1965): Computer Solutions of the Traveling Salesman Problem. *Bell System Technical J.*, 44:2245-2269

Lin, S., Kerninghan, B.W. (1973): An Effective Heuristic Algorithm for the Traveling Salesman Problem. *Operations Research*, 21:498-516

Männer, R., Manderick, B., eds. (1992): *Parallel Problem Solving from Nature II*, North-Holland, Amsterdam et al.

Maes, P. (1995): Artificial Life Meets Entertainment: Lifelike Autonomous Agents. *Communications of the ACM*, 38:108-114

Manderick, B., Weger, M., Spiessens, P. (1991): The Genetic Algorithm and the Structure of the Fitness Landscape. In Below and Boocker (eds.), pp. 143-150

Mastrolilli, M. (1998): Effective Neighborhood Functions for the Flexible Job Shop Problem. Working Paper, IDSIA Italy, Lugano

Mathias, K., Whitley, D. (1992): Genetic Operators, the Fitness Landscape and the Traveling Salesman Problem. In Männer and Manderick (eds.), pp. 219-228

Matsuo, H., Suh, C.J., Sullivan, R.S. (1988): A Controlled Search Simulated Annealing Method for the General Jobshop Scheduling Problem. Working Paper 03-04-88, Graduate School of Business, University of Texas, Austin

Mattfeld, D.C., Kopfer, H., Bierwirth, C. (1994): Control of Parallel Population Dynamics by Social-Like Behavior of GA-Individuals. In Davidor et al. (1994), pp. 16-25

Mattfeld, D.C. (1996): *Evolutionary Search and the Job Shop: Investigations on Genetic Algorithms for Production Scheduling.* Physica Verlag, Heidelberg

Mattfeld, D.C., Bierwirth, C. (1998): Minimizing Job Tardiness - Priority Rules vs. Adaptive Scheduling. In Parmee, I.C. (ed.): *Adaptive Computing in Design and Manufacture.* Springer-Verlag, London, pp. 59-67

Mattfeld, D.C., Bierwirth, C., Kopfer, H. (1999): A Search Space Analysis of the Job Shop Scheduling Problem. *Annals of Operation Research*, 86:441-453

Mattfeld, D.C., (1999): Scalable Search Spaces for Scheduling Problems. In Banzhaf, W. et al. (eds.): *Proceedings of the Genetic and Evolutionary Computation Conference.* Morgan Kaufmann Publishers, San Francisco CA, pp. 1616-1621

Maynard Smith, J. (1989): *Evolutionary Genetics.* Oxford University Press, New York

McDonald, J.R., Eberhart, R.C., eds. (1997): *Evolutionary Programming VI.* Springer Verlag, New York

Mertens, P., Falk, J., Spieck, S., Weigelt, M. (1996): Decentralized Problem Solving in Logistics with Partly Intelligent Agents and Comparison with Alternative Approaches. In König et al. (eds.), pp. 87-103

Merz, P., Freisleben, B. (1997): A Genetic Local Search Approach to the Quadratic Assignment Problem. In Bäck (ed.), pp. 465-472

Mesarović, M.D., Takahara, Y. (1975): *General Systems Theory - Mathematical Foundations.* Academic Press, New York et al.

Mesarović, M.D., Macko, D., Takahara, Y. (1970): *Theory of Hierarchical, Multilevel, Systems.* Academic Press, New York et al.

Michalewicz, Z. (1996): *Genetic Algorithms + Data Structure = Evolutionary Programs.* Springer-Verlag, New York

Michalski, R.S., Kodratoff, Y., eds. (1990): *Machine Learning.* Morgan Kaufmann Publishers, San Mateo CA

Michalski, R.S., Kodratoff, Y. (1990): Research in Machine Learning – Recent Progress, Classification of Methods, and Future Directions. In Michalski and Kodratoff (eds.), pp. 3-30

Mitchell, M., Holland, J.H., Forrest, S. (1994): When will a Genetic Algorithm outperform Hill Climbing? In Cowan, J.D., Tesauro, G., Alspector, J. (eds.): *Advances in Neural Information Processing Systems 6.* MIT Press, Cambridge MA and London

Mitchell, M. (1996): *An Introduction to Genetic Algorithms.* MIT Press, Cambridge MA

Mori, N., Kita, H., Nishikawa, Y. (1996): Adaptation to a Changing Environment by Means of the Thermodynamical Genetic Algorithm. In Voigt et al. (eds.), pp. 513-522

Morton, T.E., Pentico, D.W. (1993): *Heuristic Scheduling Systems.* John Wiley & Sons, Chichester

Mühlenbein, H., Gorges-Schleuter, M. (1988): Die Evolulutionsstrategie - Prinzip für Parallele Algorithmen. Annual Report, German National Research Center for Information Technology (GMD), Sankt Augustin

Mühlenbein, H., Gorges-Schleuter, M., Krämer, O. (1988): Evolution Algorithms in Combinatorial Optimization. *Parallel Computing*, 7:65-85

Mühlenbein, H. (1991): Evolution in Time and Space - The Parallel Genetic Algorithm. In Rawlins (ed.), pp. 316-337

Mühlenbein, H., Schlierkamp -Voosen, D. (1993): Predictive Models for the Breeder Genetic Algorithm. *Evolutionary Computation*, 1:25-49

Mühlenbein, H., Schlierkamp-Voosen, D. (1994): The Science of Breeding and its Application to the Breeder Genetic Algorithm. *Evolutionary Computation*, 1:335-360

Mühlenbein, H. (1997) Genetic Algorithms. In Aarts and Lenstra (eds.), pp. 137-171

Nakano, R., Yamada, T. (1991): Conventional Genetic Algorithms for Job Shop Problems. In Below and Boocker (eds.), pp. 474-479

Nakano, R., Davidor, Y., Yamada, T. (1994): Optimal Population Size under Constant Computational Cost. In Davidor et al. (1994), 130-138

Nilsson, N.J. (1980): *Principles of Artificial Intelligence.* Morgan Kaufmann Publishers, San Mateo CA

Nissen, V., Paul, H. (1995): A Modification of Threshold Acceptance and its Application to the Quadratic Assignment Problem. *OR Spektrum*, 17:205-210

Nowicki, E., Smutnicki,S. (1996): A Fast Taboo Search Algorithm for the Job Shop Problem. *Management Science*, 42:797-813

Oliver, L.M, Smith, D.J., Holland, J.R.C. (1987): A Study of Permutation Crossover Operators on the Traveling Salesman Problem. In Grefenstette (ed.),pp. 224-230

Osman, I.H., Potts, C.N. (1989): Simulated Annealing for Permutation Flow-Shop Scheduling. *Omega*, 17:551-557

Osman, I.H. (1993): Metastrategy Simulated Annealing and Tabu Search Algorithms for Vehicle Routing Problem. *Annals of Operations Research*, 41:421-451

Palaniswami, M., Attikiouzel, Y., Marks II, R.J., Fogel, D., Fukuda, T., eds. (1995): *Computational Intelligence - A Dynamic System Perspective.* IEEE Press, New York

Panwalkar S.S., Iskander W. (1977): A Survey of Scheduling Rules. *Operations Research*, 25:45-61

Pesch, E. (1994): *Learning in Automated Manufacturing.* Physica-Verlag, Heidelberg

Pesch, E., Voß, S. (1995): Strategies with Memories: Local Search in an Application Oriented Environment. *OR Spektrum*, 2/3:55-66

Poensgen, O.H. (1980): Koordination. In Grochla, E. (ed.): *Handwörterbuch der Organisation.* C.E. Poeschel Verlag, Stuttgart, pp. 1130-1141

Portmann, M.C. (1996): Genetic Algorithms and Scheduling: A state of the Art and some Propositions. In: *Workshop Proceedings on Production Planning and Control.* Mons, pp. 1-14

Radcliffe, N.J. (1991): Forma Analysis and Random Respectful Recombination. In Below and Boocker (eds.), pp. 222-229

Radcliffe, N.J., Surry, P.D. (1995): Fitness Variance of Formae and Performance Prediction. In Whitley and Vose (eds.), pp. 51-72

Raman, N., Rachamadugu, R., Talbot, F.B. (1989): Real-Time Scheduling of an Automated Manufacturing Center. *EJOR*, 40:222-242

Rawlins, G.J.E., ed. (1991): *Foundations of Genetic Algorithms 1.* Morgan Kaufmann Publishers, San Mateo CA

Rechenberg, I. (1973): *Evolutionsstrategie - Optimierung Technischer Systeme nach dem Vorbild der Natur.* Frommann-Holzboog, Stuttgart

Reeves, C.R., ed. (1993): *Modern Heuristic Techniques for Combinatorial Problems.* Blackwell Scientific Publications, Oxford et al.

Reeves, C.R. (1993): Genetic Algorithms. In Reeves (ed.), pp. 151-196

Reeves C. (1994): Hybrid Genetic Algorithms for Bin-packing and Related Problems. *Annals of Operations Research*, 63:371-396

Rehkugler, H., Zimmermann, H.G. (1994): *Neuronale Netze in der Ökonomie: Grundlagen und finanzwirtschaftliche Anwendungen.* Verlag Vahlen, München

Richardson, J.T., Palmer, M.R., Liepins, G.E., Hilliard, M.R.: Some Guidelines for Genetic Algorithms with Penalty Functions. In Schaffer (ed.), pp. 191-197

Rixen, I., Kopfer, H., Bierwirth, C. (1995): A Case Study of Operational Just-in-Time Scheduling using Genetic Algorithms. In Biethahn and Nissen (eds.), pp. 112-123

Rixen, I. (1997): *Maschinenbelegungsplanung mit Evolutionären Algorithmen.* DUV Gabler-Vieweg-Westdeutscher Verlag, Wiesbaden

Rochat, Y., Taillard É.D. (1995): Probabilistic Diversification and Intensification in Local Search for Vehicle Routing. *Journal of Heuristics*, 1:147-167

Rummelhart, D.E., Hintion, G.E., Williams, R.J. (1986): Learning Representations by Back-Propagation Errors. *Nature*, 323:533-536

Russell, B. (1912): *The Problems of Philosophy.* Reprint (1967), Oxford University Press, New York

Schaffer, J.D., ed. (1989): *Proceedings of the 3rd Int. Conference on Genetic Algorithms.* Morgan Kaufmann Publishers, San Mateo CA

Schneeweiß, C. (1994): Elemente einer Theorie hierarchischer Planung. *OR Spektrum*, 16:161-168

Schneeweiß, C. (1995): Hierarchical Structures in Organizations: A Conceptual Framework. *EJOR*, 86:4-131

Schwefel, H.-P. (1977): *Numerische Optimierung von Computermodellen mittels der Evolutionsstrategie.* Birkhäuser, Basel

Schwefel, H.-P. (1995): *Evolution and Optimum Seeking.* John Wiley, New York

Schwefel, H.-P., Männer, R., eds. (1990): *Parallel Problem Solving from Nature I.* Springer-Verlag, Heidelberg

Smith, R.G. (1980): The Contract Net Protocol: High-Level Communication and Control in a Distributed Problem Solver. *IEEE Transactions on Computers C*, 29:1104-1113

Smith, D. (1985): Bin Packing with Adaptive Search. In Grefenstette (ed.), pp. 165-170

Stadler, P.F., Schnabl, W. (1992): The Landscape of the Traveling Salesman Problem. *Physics Letters A*, 161:337-344

Stadler, P.F., Happel, R. (1992): Correlation Structure of the Landscape of the Graph Bi-Partitioning Problem. *Phys. A: Math. Gen.*, 25:3103-3110

Stadtler, H. (1988): *Hierarchische Produktionsplanung bei losweiser Fertigung.* Physica-Verlag, Heidelberg

Steinhöfel, K., Albrecht, A., Wong, C.K. (1998): Two Simulated Annealing-based Heuristics for the Job Shop Scheduling Problem. Working Paper, German National Research Center for Information Technology (GMD), Berlin

Stöppler, S., Bierwirth, C. (1992): The Application of a Parallel Genetic Algorithm to the $n/m/P/C_{max}$ Problem. In Fandel, G., Gulledge, T., Jones, A. (eds.): *New Directions for Operations Research in Manufacturing.* Springer-Verlag, Berlin et al., pp. 161-175

Storer, R., Wu, D., Vaccari, R. (1992): New Search Spaces for Sequencing Problems with Application to Job Shop Scheduling. *Management Science*, 38:1495-1509.

Swinburne R. (1974) (ed.): *The Justification of Induction.* Oxford University Press, New York et al.

Syswerda, G. (1989): Uniform Crossover in Genetic Algorithms. In Schaffer (ed.), pp. 2-9

Syswerda, G. (1991): Schedule Optimization using Genetic Algorithms. In Davis (ed.), pp. 332-349

Taillard, É. (1993): Benchmarks for Basic Scheduling Problems. *EJOR*, 64:278-285

Taillard, É.D. (1994): Parallel Taboo Search Techniques for the Job Shop Scheduling Problem. *ORSA Journal on Computing*, 6:108-117

Taillard, É.D. (1995): Comparison of Iterative Searches for the Quadratic Assignment Problem. *Location Science*, 3:87-105

Tanese, R. (1989): Distributed Genetic Algorithms. In Schaffer (ed.), pp. 434-439

Tempelmeier, H., Helber, S. (1994): A Heuristic for Dynamic Multi-Item Multi-Level Capacitated Lotsizing for General Product Structures. *EJOR*, 75:296-311

Thrift, P (1991): Fuzzy Logic Synthesis with Genetic Algorithm. In Below and Boocker (eds.), pp. 509-513

Tsujimura, Y., Cheng, R., Gen, M. (1997): Improved Genetic Algorithms for Job-Shop Scheduling Problems. *Engineering Design & Automation*, 97:133-144

Ulder, N.L.J., Aarts, E.H.L., Bandelt, H.-J., Laarhoven, P.J.M., Pesch, E. (1990): Genetic Local Search Algorithms for the Traveling Salesman Problem. In Schwefel and Männer (eds.), pp. 109-116

Unland, R., Kirn, S., Wanka, U., O'Hara, G.M.P., Abbas, S. (1996): Organizational Multi-Agent Systems: A Process Driven Approach. In König et al. (eds.), pp. 105-122

Vaessens, R.J.M, Aarts, E.H.L., Lenstra, J.K. (1992): A Local Search Template. In Männer and Manderick (eds.), pp. 65-74

Vaessens, R.J.M., (1995): *Generalized Job Shop Scheduling: Complexity and Local Search.* Ph.D. Thesis, Eindhoven University of Technology, Eindhoven

Van Dyke Parunak, H. (1992): Characterizing the Manufacturing Scheduling Problem. *Journal of Manufacturing Systems*, 69:187-199

Van Dyke Parunak, H., Fulkerson, B. (1994): Genetic Algorithms and Production Scheduling. *GA Digest* 8(8), FTP.AIC.NRL.Navy.Mil

Vavak, F., Fogarty, T., Jukes, K. (1996): A Genetic Algorithm with Variable Range of Local Search for Tracking Changing Environments. In Voigt et al. (eds.), pp. 376-385

Vegte, J. van de (1990): *Feedback Control Systems.* Prentice Hall, Englewood Cliffs

Voigt, H.-M., Ebeling, W., Rechenberg, I., Schwefel, H.-P., eds. (1996): *Parallel Problem Solving from Nature IV,* Springer-Verlag, Berlin

Vose, M., Liepins, G.E. (1991): Punctuated Equilibria in Genetic Search. *Complex Systems*, 5:31-44

Wagner, H.M., Whitin, T.M. (1958): Dynamic Version of the Economic Lot Size Model. *Management Science*, 5:145-156

Weinberger, E. (1990): Correlated and Uncorrelated Fitness Landscapes and How to Tell the Difference. *Biological Cybernetics*, 63:325-336

White, K.P., Rogers, R.V. (1990): Job Shop Scheduling: Limits of the Binary Disjunctive Representation.. *Int. Journal of Production Research*, 28:2187-2200

Whitley, D. (1989): The Genitor Algorithm and Selection Pressure. In Schaffer (ed.), pp. 116-121

Whitley, D., Domonic, S., Das, R. (1991): Genetic Reinforcement Learning with Multilayer Neural Networks. In Below and Boocker (eds.), pp. 562-569

Whitley, L.D., ed. (1993): *Foundations of Genetic Algorithms 2.* Morgan Kaufmann Publishers, San Francisco CA

Whitley, L.D., Vose, M., ed. (1995): *Foundations of Genetic Algorithms 3.* Morgan Kaufmann Publishers, San Francisco CA

Wolpert, D.H., Macready, W.G. (1997): No Free Lunch Theorems for Search. *IEEE Transactions on Evolutionary Computation*, 1:67-82

Yamada, T., Nakano, R. (1992): A Genetic Algorithm Applicable to Large-Scale Job Shop Problems. In Männer and Manderick (eds.), pp. 281-290

Zelewski, S. (1993): Multi-Agenten-Systeme für die Prozekoordinierung in komplexen Produktionssystemen. Industrial Production Economics Report No. 46, Dept. of the University of Cologne, Cologne

Zimmermann, H.J. (1987): *Fuzzy Sets, Decision Making and Expert Systems.* Kluwer Academic Press, Boston et al.

Index

MIX
Papier aus verantwortungsvollen Quellen
Paper from responsible sources
FSC® C105338

If you have any concerns about our products,
you can contact us on
ProductSafety@springernature.com

In case Publisher is established outside the EU,
the EU authorized representative is:
Springer Nature Customer Service Center GmbH
Europaplatz 3, 69115 Heidelberg, Germany

Printed by Libri Plureos GmbH
in Hamburg, Germany